Issued under the authority of the
(Fire and Emergency Planning Di

Fire Service Manual

Volume 1
Fire Service Technology, Equipment and Media

Hydraulics, Pumps and Water Supplies

HM Fire Service Inspectorate Publications Section
London: The Stationery Office

ISBN 0 11 341216 9

Cover photograph: Shropshire Fire and Rescue Service

Half-title page photograph: Shropshire Fire and Rescue Service

Printed in the United Kingdom by The Stationery Office
TJ4519 5/01 C50 019585

Hydraulics, Pumps and Water Supplies

Preface

This new Fire Service Manual, 'Hydraulics, Pumps and Water Supplies', is designed to replace Book 7 of the Manual of Firemanship, 'Hydraulics, pumps and pump operation' and much of the text is taken from that publication because it remains relevant today. However, the format of this new Manual differs significantly from that of its predecessor.

The hydraulics theory has been condensed to that which is considered necessary for firefighters to:

- understand the behaviour of water and the firefighting equipment which controls it.
- make informed decisions regarding the supply and management of water both on the fireground and when pre-planning.

In order to maintain the flow of the main text, derivations of the hydraulics formulae have been removed from it, but those readers who wish to study them in detail will find them in the appendices section.

Additionally, there is no longer a requirement for a knowledge of logarithms and, following consultation with brigades, it became clear that the material in the appendices on graphs and on powers and roots could, if required, best be studied in a mathematics text book. Consequently these sections have been removed.

Some of the most significant developments since Book 7 was published have been:

(i) The obligations placed upon the privatised water undertakers to reduce leakage have resulted in lower mains operating pressures which, in turn, have created problems regarding water for firefighting.

(ii) The increasing use made by brigades of large diameter hose and the means to deploy and retrieve it.

(iii) Initiatives regarding the development of flowmetering and of automatic pump and tank fill controls have come to fruition, though not all brigades are taking advantage of such equipment.

(iv) A greater emphasis on safe working practices.

(v) The substantial replacement of the type "A" branch with more modern, adjustable branches. Reference to the former has, however, been retained in this publication.

This new publication has been written giving due regard to these developments.

Home Office, 2001

Hydraulics, Pumps and Water Supplies

Contents

Hydraulics, Pumps and Water Supplies

Hydraulics, Pumps and Water Supplies

Chapter

1

Chapter 1 – Elementary Principles of Hydraulics

Introduction

The term hydraulics refers to the study of the behaviour of water both when it is in motion and when it is at rest. A grasp of the elementary principles of the subject is necessary to enable firefighters to:

(i) understand the behaviour of water in relation to the operation of fire service equipment.

(ii) make informed decisions concerning the supply and delivery of water in the dynamic arena of the fireground.

(iii) make effective arrangements for the provision of water supplies at the pre-planning stage.

1.1 The Properties of Water

Water when pure is a colourless, odourless liquid with a molecular composition of two atoms of hydrogen combined with one atom of oxygen (H_2O). A litre of water has a mass of 1 kilogram (kg), and a cubic metre, which is for all practical purposes 1000 litres, therefore has a mass of 1000kg or 1 tonne.

The weight (i.e. the downward force which gravity exerts) of a body depends on its mass (m), and also on how powerfully it is attracted by gravity. It is shown in the Fire Service Manual Volume 1 'Chemistry and Physics for Firefighters' that:

weight = mg newtons (N)

where g is the acceleration due to gravity. The value of g varies very slightly over the surface of the earth but, to a close approximation, it is 9.81 metres per second per second ($9.81 m/s^2$), so the weight of 1 kilogram of water is

$$1 \times 9.81 = 9.81N$$

A cubic metre of water therefore exerts a downward force of 9810 N (1000 × 9.81), or 9.81 kilonewtons (kN).

For most practical purposes therefore:

1 litre of water weighs 10 newtons
1 cubic metre of water weighs 10 000 newtons

Although the density (see Fire Service Manual Volume 1 'Chemistry and Physics for Firefighters') of water varies with the degree of purity, such variations are very small and may usually be ignored. Sea water, for example, has a density of approximately 1.03 kg/litre

Pure water has a freezing point of 0°C and a boiling point of 100°C, both at normal atmospheric pressure (1 bar approx.). Between these temperatures at atmospheric pressure, therefore, water exists as a liquid, and exhibits all the characteristic properties of a fluid. It is virtually incompressible, and an increase of 1 bar only causes a decrease in volume of 0.000 002 per cent.

As a fluid, water has volume but is incapable of resisting change of shape, i.e. when poured into a container it will adjust itself irrespective of the shape of the latter, and will come to rest with a level surface. This is because there is very little friction or cohesion between the individual molecules of which water is composed.

Water, of the degree of purity likely to be used for firefighting purposes, is a reasonably good conductor of electricity and therefore great care should be taken to prevent firefighting streams from coming into contact with live electrical equipment. (See Fire Service Manual Volume 2 'Fire Service Operations – Electricity'.)

1.2 Principal characteristics of pressure

The S.I. unit of pressure is the newton per square metre (N/m^2) another name for which is the Pascal. However, it has been decided that, because this is a very small unit, the fire service unit of pressure will be the 'bar'. The relationship between these units is:

1 bar = 100 000 N/m^2 or 10^5 N/m^2

Normal atmospheric pressure = 1.013 bar.

There are a number of basic rules governing the principal characteristics of pressure in liquids. These are:

(a) The pressure exerted by a fluid at rest is always at right angles to the surface of the vessel which contains it.

(b) The pressure at any point in a fluid at rest is the same in all directions.

A gauge connected onto a line of piping or hose, or to the bottom of a storage tank, containing a fluid at rest will give the same reading no matter what its orientation because the pressure exerted by the fluid is the same in all directions. This is equally true regardless of whether the pressure is due to the height of the column of fluid itself or is externally applied e.g. by a pump.

(c) Downward pressure of a fluid in an open vessel is proportional to its depth.

Figure 1.1 shows three vertical containers, The depth of water in them is 10m, 20m and 30m. If pressure gauges were to be placed at the bottom of each container they would show readings of approximately 1 bar, 2 bar and 3 bar respectively i.e. the pressure indications would be in the same ratio as the depths.

(d) The downward pressure of a fluid in an open vessel is proportional to the density of the fluid.

In Figure 1.2 are shown two containers, one holding mercury and the other water. The depth of liquid is the same in both containers. If pressure gauges were to be placed at the bottom of each container the pressure at the bottom of the mercury container would be found to be 13.6 times the pressure at the bottom of the water container because mercury is 13.6 times as dense as water.

(e) The downward pressure of a fluid on the bottom of a vessel is independent of the shape of that vessel.

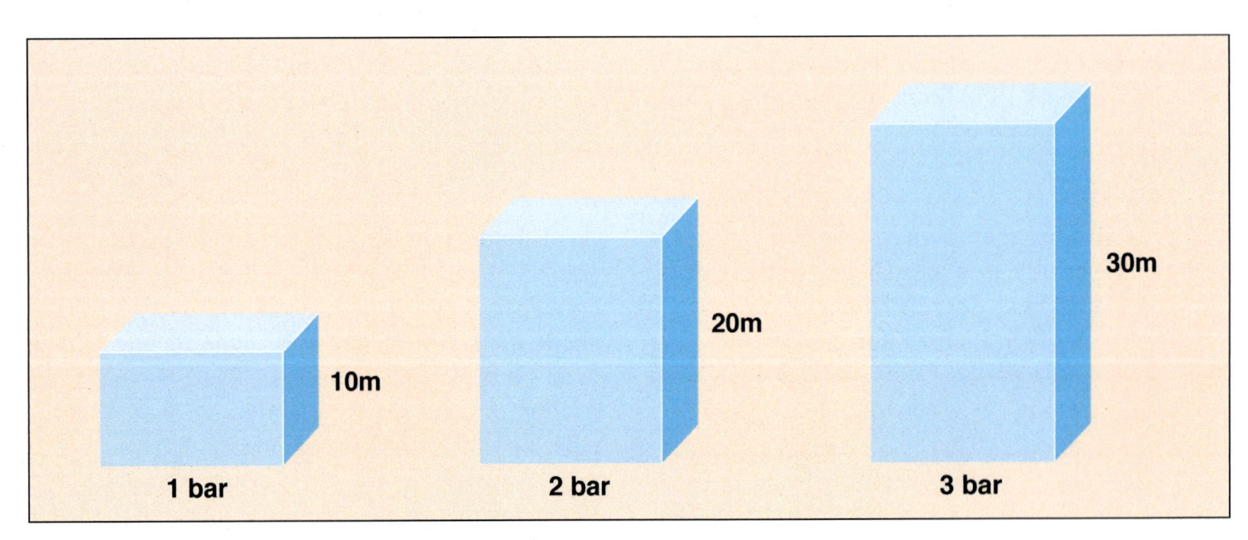

Figure 1.1(c) Downward pressure of a fluid in an open vessel is proportional to its depth.

Figure 1.2 (d) Downward pressure of a fluid in an open vessel is proportional to the density of the fluid

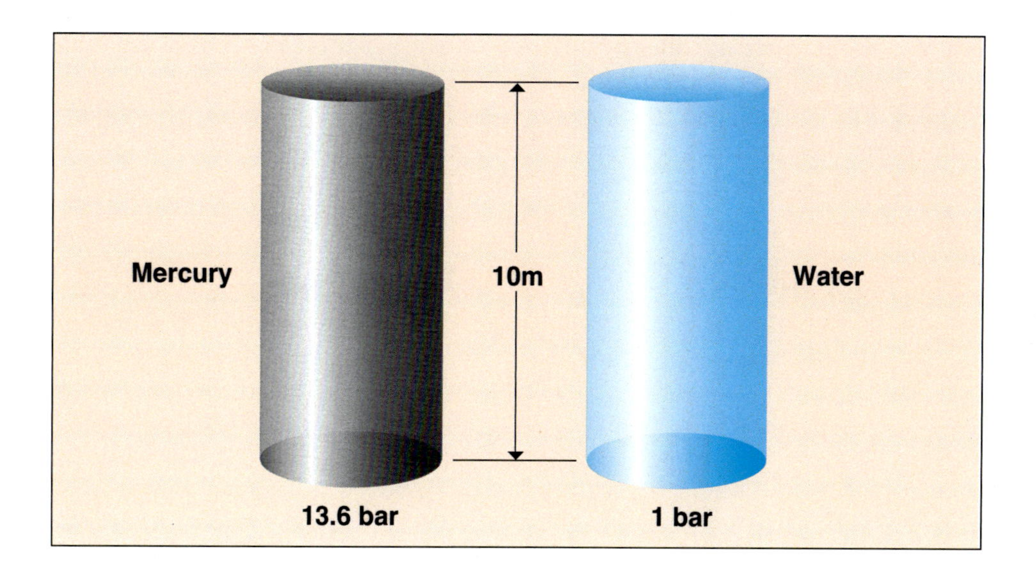

This last principle is illustrated in Figure 1.3, which shows a number of containers of varying shapes. The pressure at the bottom of each is exactly the same provided the depth of liquid (or head) is the same in each case.

One practical consequence of this principle is that the static (i.e. no flow) pressure at the delivery end of a pipeline leading from an elevated storage tank or reservoir is decided solely by the vertical distance between the water surface and the point where the pressure is measured, because the head of water is the same no matter how convoluted the route which the pipeline takes. Also, if water is being pumped up to such a container, or to an elevated branch for firefighting purposes, the amount of pump pressure required to overcome the head will be independent of the precise route which the pipe or hose takes. However, once flow is taking place, friction between the moving water and the inside surface of the pipe or hose becomes a complicating factor which will be considered later in this chapter.

1.3 Relationship between Pressure and Head for Water

It has already been stated that the pressure of a liquid contained in an open vessel is proportional both to the depth of the liquid and to its density. The precise relationship for pressure is:

$$p = H\rho g \qquad \text{(see Appendix 4 for derivation)}$$

Figure 1.3 (e) The downward pressure of a fluid on the bottom of a vessel is independent of the shape of that vessel.

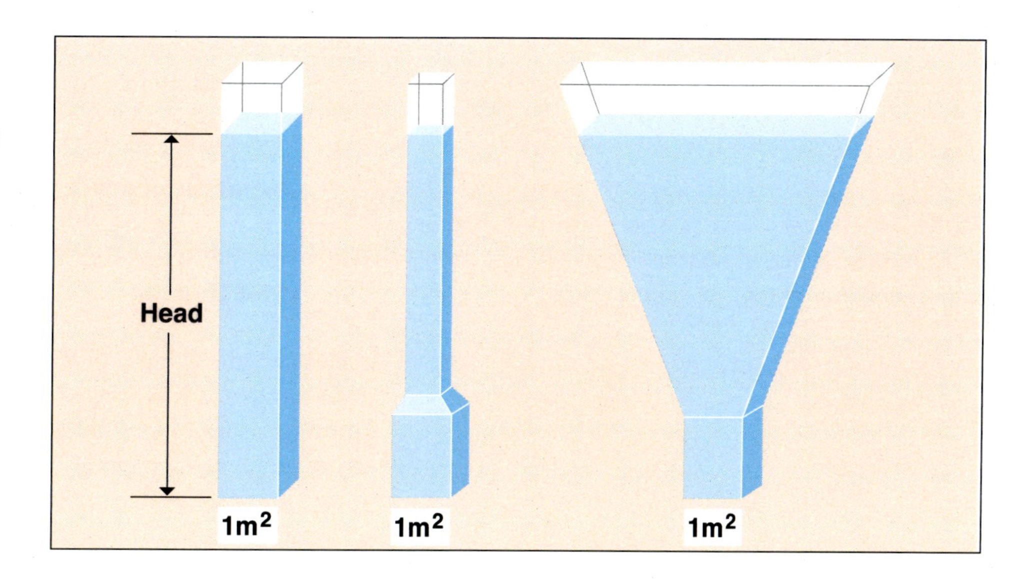

where p is the pressure in newtons per square metre, H is the head (depth) of liquid in metres, ρ is the density of the liquid in kilograms per cubic metre (kg/m^3) and g is the acceleration due to gravity. Taking ρ as 1000 kg/m^3 and g as 9.81 m/s^2 gives

$$p = H \times 1000 \times 9.81 \quad N/m^2$$

$$\mathbf{P = 0.0981 \times H} \quad \textbf{bar}$$

Transposing this formula to find H we have:

$$H = \frac{P}{0.0981} \quad \text{metres}$$

or $$\mathbf{H = 10.19 \times P} \quad \textbf{metres}$$

To a close approximation these two formulae simplify to:

$$\mathbf{P = \frac{H}{10}} \quad \textbf{bar}$$

$$\mathbf{H = 10 \times P} \quad \textbf{metres}$$

Example 1

A static water pressure of 6 bar is required at a certain point from a supply to be obtained from an elevated reservoir. At what vertical height above this point must the water surface be?

$$H = 10 \times P \quad \text{i.e. } 10 \times 6$$

$$= 60 \text{ metres head}$$

(The exact formula gives 61.14 metres head)

Example 2

The water level of a tank in a sprinkler installation is 40 metres above a pressure gauge. What pressure in bar should register on the gauge?

$$P = \frac{H}{10} \quad \text{i.e. } \frac{40}{10}$$

$$= 4 \text{ bar}$$

(The exact formula gives 3.924 bar).

It is clear from these examples that the error introduced by using the approximate formulae is small and likely to be within the limits of accuracy of most gauges.

Example 3

A pump pressure of 5 bar is required to operate a branch when it is working at ground level. If the branch is raised to a height of 25m above the ground what must be the pump pressure in order to maintain the same output?

The additional pressure required to overcome the head of 25m is given by:

$$P = \frac{H}{10} \quad \text{i.e. } \frac{25}{10}$$

$$= 2.5 \text{ bar}$$

Thus the new pump pressure will need to be:

$$5 + 2.5 = 7.5 \text{ bar}$$

1.4 Loss of Pressure due to Friction

When water flows through a hose or pipe there is a gradual loss of pressure resulting from the need to overcome the frictional resistance which exists between the moving water stream and the internal surface of the hose or pipe. The magnitude of this frictional resistance depends on the various factors discussed in paragraph 1.4.1. It is of particular significance:

(i) when water is drawn from hydrants on small diameter mains;

(ii) when pumping water over long distances by means of a relay;

(iii) on the fireground when small diameter hose is used.

The accurate estimation of friction loss by calculation is notoriously difficult but, fortunately, this is seldom necessary. What is more important is to appreciate which physical factors are most significant in determining friction loss so that due allowance may be made for them in the types of situations listed above. Such practical considerations are discussed in detail in later chapters.

1.4.1 Laws governing loss of pressure due to friction

Experiments on the flow of water through hose and pipes show that, to a reasonably good approximation, the loss of pressure (P_f) due to friction is governed by the following laws:

(i) P_f is directly proportional to the length (l) of hose through which the water flows.

i.e. $P_f \propto l$

Thus, if 0.5 bar pressure is lost when water flows through 1 hose length, then 5 bar will be lost if the same quantity of water flows through 10 lengths.

(ii) P_f is directly proportional to a quantity called the friction factor (f) for the hose (determined largely by the roughness of its inside surface).

i.e. $P_f \propto f$

The friction factor is also affected by other influences, in particular whether the couplings are smaller or larger than the hose itself, but the values quoted normally allow for these factors.

Table 1.1 below gives the approximate friction factors for a variety of hose diameters, but it should be appreciated that different samples of hose, although of nominally the same diameter, may take slightly different values.

Table 1.1 Friction factors for hose

Diameter of hose	Friction factor
38mm, 45mm, 64mm and 70mm	0.005
90mm with standard instantaneous couplings	0.007
90mm with full flow couplings	0.005
100mm and 125mm	0.004
150mm	0.003

(iii) P_f is directly proportional to the square of the flowrate (l).

i.e. $P_f \propto L^2$

Thus, for example, if the flowrate through a line of hose is doubled, the pressure loss due to friction will be increased by a factor of four.

(Since, for a given pipe diameter, velocity of flow is directly proportional to flowrate, an equivalent statement to this law is that friction loss is proportional to the square of the velocity of the water.)

(iv) P_f is inversely proportional to the fifth power of the hose diameter (d).

i.e. $P_f \propto \frac{1}{d^5}$

Diameter is the most important single factor which affects friction loss. Because of the fifth power law, a modest change in hose diameter produces a dramatic change in friction loss. Taking what is admittedly an extreme case, the friction loss in 45mm hose is approximately 32 (2^5) times greater than the friction loss in 90mm hose with full flow couplings, for the same flowrate. This extreme change is largely due to the fact that, in the larger hose, because the area of cross section is four times as great, the velocity of flow is only a quarter of that in the smaller hose.

1.4.2 The Friction Loss Formula

There are several formulae which may be used for the calculation of friction loss and the greater the required degree of accuracy required the more complex the formula becomes. For firefighting situations, where the highest accuracy is not required, the most useful relationship brings together the four proportionality statements given above and can be shown to be:

$$P_f = \frac{9000 f l L^2}{d^5}$$

in which **f** is the friction factor, P_f the pressure loss in **bars**, l the length of hose in **metres**, **L** the flowrate in **litres per minute** and **d** the hose diameter in **millimetres**.

Example 4

The flowrate in 50m of 45mm diameter hose is 400 litres per minute (l/min). What is the loss of pressure in the hose if the friction factor is 0.005?

$$P_f = \frac{9000 f l L^2}{d^5}$$

$$\text{i.e. } P_f = \frac{9000 \times 0.005 \times 50 \times 400 \times 400}{45 \times 45 \times 45 \times 45 \times 45}$$

i.e. the loss
of pressure = 2.0 bars (or 1 bar per length)

Example 5

Calculate the pressure loss due to friction if:

(i) 90mm hose with a friction factor of 0.007

(ii) twin lines of 45mm hose

were to be substituted for the 45mm hose in Example 4.

(i) for the 90mm hose:

$$P_f = \frac{9000 \times 0.007 \times 50 \times 400 \times 400}{90 \times 90 \times 90 \times 90 \times 90}$$

i.e. the loss
of pressure = 0.09 bar (approximately 0.05 bar per length)

(ii) for the twinned 45mm hose the flowrate in each line is 200 l/min so:

$$P_f = \frac{9000 \times 0.005 \times 50 \times 200 \times 200}{45 \times 45 \times 45 \times 45 \times 45}$$

= 0.5 bar (or 0.25 bar per length)

The advantage of using large diameter hose in relay and fireground situations is clearly illustrated by these examples, though in the latter case, ease of handling and manoeuvrability of hose lines may be more important.

Table 6.1 in Chapter 6 gives approximate pressure losses due to friction for various hose diameters and flowrates.

1.5 Energy Changes in Water Streams

When water flows through channels of varying cross section, such as nozzles, couplings, venturi, and the volutes of pumps, changes in velocity and pressure occur. Bernoulli's equation, which is derived in Appendix 4, gives the precise quantitative relationship between these variables, but an understanding of the observed behaviour of the water may be achieved by considering the changes in the various forms of energy which occur as the water flows through the channel. These forms of energy, most of which are explained in the Fire Service Manual Volume 1 – 'Physics and Chemistry for Firefighters' are:

(i) Kinetic – energy which the water has because of its velocity

(ii) Potential – energy which the water has because of its height above a fixed reference point such as an outlet or pump

(iii) Pressure

(iv) Heat

Application of the Principle of the Conservation of Energy, which dictates that energy cannot be created or destroyed but only transformed from one form into another, implies that if one form of energy is increased then that increase will be reflected by a decrease in one or more of the other forms. This principle applies to water flowing in the types of channel mentioned in the following paragraphs.

Because of friction, heat energy is created at the expense of pressure when water flows through a line of hose, though it would be difficult to detect any increase in the temperature of the water. Once heat energy has been produced it cannot easily be converted back to any of the other forms. Heat energy is also created at any points in a water stream where turbulence is caused by sudden changes in cross section or the direction of flow. For most practical purposes temperature changes in water streams may be ignored though when a centrifugal pump is running against a closed, or very restricted, delivery the water gets noticeably hot because much of the energy supplied to the pump is wasted in the creation of turbulence.

1.5.1 Flow through Nozzles

Figure 1.4 shows water flowing through a Type A nozzle. As the water flows from point A to point B its velocity, and hence its kinetic energy, increase but only at the expense of the pressure which decreases from a few bar at point A to atmospheric at point B. There are no significant changes in the other types of energy.

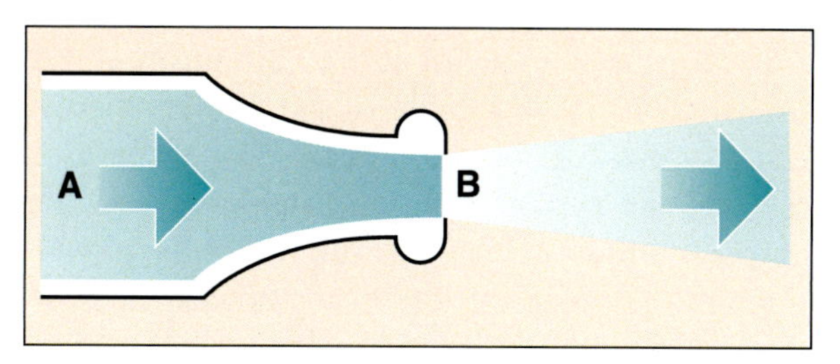

Figure 1.4 Water flowing through a Type A nozzle.

It is shown in Appendix 4 that the relationship between the velocity of the jet, **v**, and the original pressure, **P**, is:

$$\mathbf{v} = \mathbf{14.14\sqrt{P}}$$

and that the number of litres per minute (l/min) discharged, **L**, is given by:

$$\mathbf{L} = \mathbf{\frac{2}{3}d^2\sqrt{P}}$$

Example 6

Calculate the flowrate from a 15mm type A nozzle when working at a pressure of 4 bar.

$$L = \frac{2}{3}d^2\sqrt{P}$$

$$\text{flowrate} = \frac{2}{3} \times 15 \times 15 \times \sqrt{4}$$

$$= 300 \text{ litres per minute}$$

N.B. this formula cannot be applied to diffuser and jet/spray type nozzles because the water is not discharged through a simple circular section.

1.5.2 Flow through a Venturi

Figure 1.5 shows water flowing through a venturi, i.e. a section of pipe in which the diameter gradually reduces from its initial value, at point A, to a minimum, at the throat, B, before increasing again. Because the changes in diameter are gradual, little turbulence is created in the water stream.

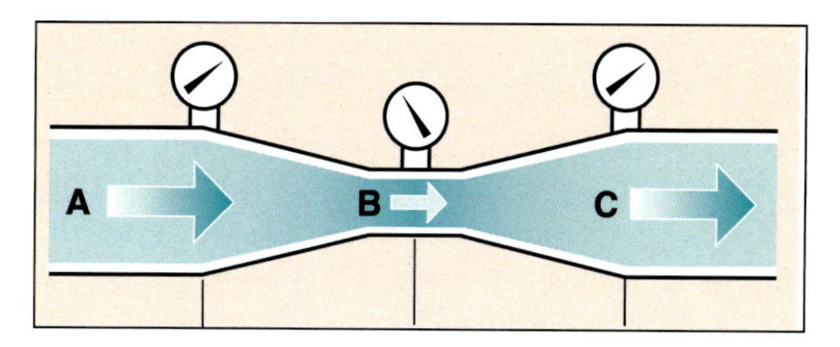

Figure 1.5 Water flowing through a venturi.

As the water flows from A to B its velocity, and hence its kinetic energy, increases to a maximum at the expense of pressure which falls to a minimum value at B. At point C the kinetic energy has decreased to its initial value (if the diameters at A and C are the same) and the pressure recovers to close to its former value. If the pressure at C is not too high, it is possible for the pressure at the throat of the venturi to fall to below that of the atmosphere with the result that air, water or any other fluid outside the device will be drawn (induced) into the stream through any opening which may exist at the throat. This is the principle underlying the operation of some types of flowmeter, the foam inductor and of the ejector pump which is described in Chapter 5.

1.5.3 Flow through Sudden Reductions in Diameter

Figure 1.6 shows water flowing through a channel in which the diameter abruptly reduces from its initial value, at point A, to a smaller value, at point B before abruptly increasing again.

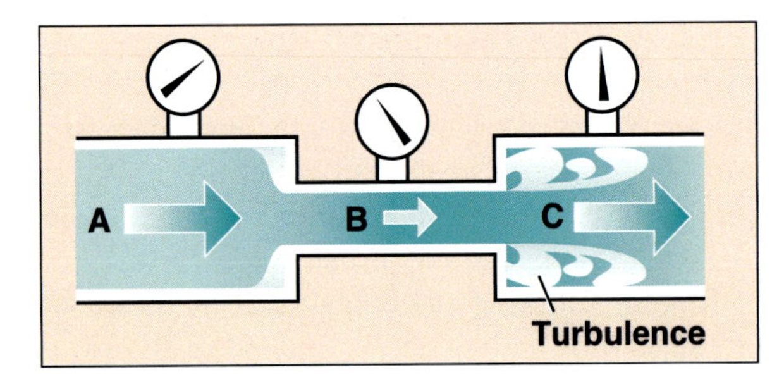

Figure 1.6 Water flowing through a channel having a sudden change in diameter

As with the venturi, there is an increase in kinetic energy as the water is forced through the reduced cross section and a consequential reduction in pressure. However, although the kinetic energy reduces to its initial value as the water stream returns to its original diameter, at point C, there is by no means a complete recovery in pressure. This is because the sudden change in diameter causes turbulence, particularly on the downstream side of the restriction, and some of the pressure energy is irrecoverably converted to heat energy.

Standard instantaneous couplings used with any hose diameter above 64mm, kinks in a line of hose, and mains water meters of smaller diameter than the pipes in which they are installed, all cause turbulence and a corresponding overall loss of pressure. The more dramatic the change in cross section, the greater will be the penalty in terms of pressure loss.

1.5.4 Flow through a Single Stage Pump

Figure 1.7 shows the changes in pressure and velocity (and therefore kinetic energy) which occur as water is pumped, through a single stage pump, from an open source to a jet on the fireground.

Tracing the velocity line from (1) to (6) it will be seen that the velocity through the suction hose from (1) to (2) is constant and comparatively low because of the large diameter, but there is a small loss of pressure energy mainly to compensate for the increase in potential energy as the water is lifted. After reaching the entry to the impeller at (2) the velocity and pressure *both* increase until the water leaves the impeller at (3). This apparent contradiction of the principle of the conservation of energy can occur because, at this stage, energy is fed into the system from an external source (the engine). From (3) to (4) the water is passing through the volute or diffuser, gradually decreasing in velocity until it reaches (4) the outlet of the pump casing and the inlet of the delivery hose. The decrease in kinetic energy is accompanied by a corresponding increase in pressure. From (4) to (5) the water is passing through the delivery hose and maintains a constant velocity though there is a gradual reduction of pressure due to friction loss. Between (5), the end of the delivery hose and the entry to the branch, and (6), the outlet of the nozzle, the kinetic energy increases at the expense of pressure energy as explained previously.

The energy changes which take place when a two stage pump is used are similar to those described above except that stages (2) to (4) are repeated because of the two impellers.

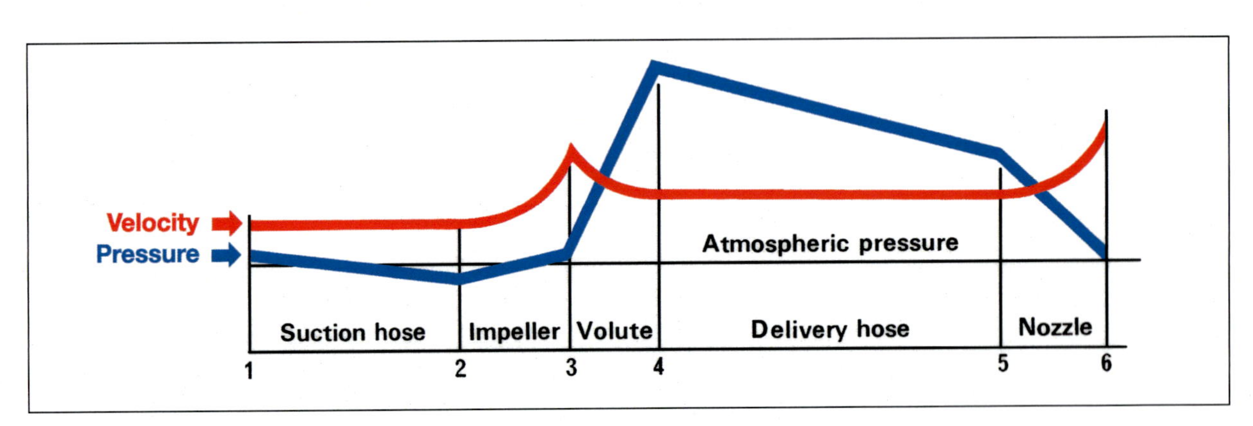

Figure 1.7 Changes in pressure and velocity as water is pumped from a supply to the nozzle.

1.6 Water Power and Efficiency

When water passes through a pump its total energy is increased because energy is supplied from an external source, i.e. the engine which drives the pump. As the kinetic energy of the water is small, both on entering and leaving the pump, the newly acquired energy is almost entirely in the form of pressure. It is shown in Appendix 4 that the water power (i.e. the energy created per second) produced by a pump is given by the formula:

$$\mathbf{WP} = \frac{\mathbf{100LP}}{\mathbf{60}}$$

where WP is the water power measured in watts, L the flowrate measured in litres per minute (l/min) and P the increase in pressure, between inlet and outlet, measured in bar.

Example 7

Calculate the water power of a portable pump which, at full power, delivers 1600 l/min when increasing the pressure by 9 bar.

$$\text{WP} = \frac{100 \times 1600 \times 9}{60} \quad \text{watts}$$

$$= 24\ 000 \text{ watts}$$

i.e. Water Power = 24 kilowatts (kW)

Some of the energy supplied to drive the pump will, because of internal turbulence, be converted to heat so it is therefore not 100% efficient in its ability to convert the brake power, BP, supplied by the engine or motor, into water power.

The efficiency of any machine is defined as:

$$\textbf{efficiency} = \frac{\textbf{useful power delivered by the machine}}{\textbf{power required to drive the machine}}$$

i.e., for a pump:

$$\textbf{efficiency} = \frac{\mathbf{WP}}{\mathbf{BP}}$$

Because it is more convenient to express efficiency, E, in percentage terms, this formula is usually written as:

$$\mathbf{E} = \frac{\mathbf{WP \times 100}}{\mathbf{BP}}$$

It is important, when using the formula, to appreciate that WP and BP should be measured in the same units, i.e. both in watts or both in kilowatts.

Example 8

The manufacturer's technical data for the portable pump in Example 7 indicates that the rated power of the engine is 37 kW. What is the efficiency of the pump?

$$\text{E} = \frac{24}{37} \times 100$$

i.e. efficiency = 65% approximately

Example 9

A pump is required to deliver 2700 l/min at 10 bar. Assuming an efficiency of 65%, calculate the brake power required to drive it.

$$\text{WP} = \frac{100 \times 2700 \times 10}{60}$$

i.e. Water Power = 45 000 watts or 45 kW

Transposing the formula for E to make BP the subject gives:

$$\text{BP} = \frac{\text{WP} \times 100}{\text{E}}$$

$$= \frac{45 \times 100}{65}$$

i.e. Brake Power = 69 kW approximately

It should be appreciated that the efficiency of a pump is by no means a fixed quantity but varies widely according to the operating conditions. For example, if the pump is operating against a closed delivery, the efficiency is zero because all the energy supplied is converted to heat. It is usually highest when the pump is operating at, or close to, its quoted duty point.

1.7 Jet Reaction

When water is projected from a nozzle, a reaction equal and opposite to the force required to discharge the jet takes place at the nozzle which tends to recoil in the opposite direction to the flow. Thus the firefighter(s), or whatever is supporting the branch, must be prepared for and be capable of absorbing this reaction.

At least one fatality and many injuries have been caused by firefighters' inability to cope with unexpectedly large jet reactions.

The whole of the reaction takes place as the water leaves the nozzle, and whether or not the jet strikes a nearby object has no effect on the reaction, though the object itself will experience a force of similar magnitude. Thus, whether or not a jet held by a firefighter on a ladder strikes a wall is immaterial to his stability on the ladder, which is governed solely by the reaction at the nozzle.

It is shown in Appendix 4 that, for a Type A nozzle, the reaction, R, measured in newtons, is given by:

$$\mathbf{R = 0.157\ P\ d^2}$$

where P is the nozzle pressure in bars and d is its diameter in millimetres

Whilst it is often possible for one firefighter to hold a small jet, several may be required for a large jet, even though the operating pressure of both may be the same, because the reaction depends not only on the velocity of the jet but also on the mass of water discharged per second. For large jets some form of support, such as a branch holder, may well be required.

Example 10

What is the difference between the reaction of the water leaving a 25mm nozzle as compared with a 12.5mm nozzle if the pressure in both cases is 7 bar?

(a) With 25mm nozzle:

$$R = 0.157 \times 7 \times 25^2$$

$$= 687 \text{ newtons.}$$

(b) With 12.5mm nozzle:

$$R = 0.157 \times 7 \times 12.5^2$$

$$= 172 \text{ newtons.}$$

The difference between the two reactions is therefore approx. 515 newtons.

Advantage can be taken of jet reaction to drive a fireboat without propelling machinery, solely by jets fixed at the rear of the vessel. Water from the fire pumps can be by-passed through these jets, and the reaction of the water leaving the nozzles propels the boat in the opposite direction to the reaction from the jets. No increase in the speed of the vessel is gained by the jets striking the surface of the water, or by their being placed under water; the reaction causing the propulsion of the boat takes place entirely at the nozzle. This can be demonstrated by diverting the water from underwater propelling jets to the monitor, when the boat will travel at approximately the same speed.

1.8 Water Hammer

This is a phenomenon with which most people will be familiar because it frequently occurs in domestic situations such as when the flow of water through a long run of metal pipe is stopped very quickly by the rapid closing of a tap. The consequent metallic 'clunk' which may be heard is the consequence of 'water hammer'.

When a moving object, such as a vehicle or a column of water, undergoes a change in velocity the force (F) required to accelerate or decelerate it depends on its mass (m), its velocity (v) and the time (t) in which the change in velocity takes place. Appendix 4 shows that the relationship between these quantities is:

$$\mathbf{F = \frac{mv}{t}}$$

where F is measured in newtons, m in kilograms, v in metres per second and t in seconds. The product $m \times v$ is known as the *momentum* of the moving object and the force, F, is therefore equal to the change of momentum per second.

One of the most important implications of the formula is that the force required to bring an object to rest depends inversely on the braking time – the shorter the time the greater the force exerted. Thus, if a vehicle is brought to rest in 0.1seconds as the result of a collision the braking force will be 100 times greater than if it is brought to rest in, say, 10 seconds as a result of normal braking, and the effect of this large force on the vehicle (and on the object with which it collides) will be only too evident!

There are a number of fireground situations where the time taken to terminate the flow of a substantial mass of water, moving with considerable velocity, may be very short and where, as a consequence, damage to equipment may result from the very large forces involved. Such situations include:

(i) shutting down a branch rapidly

Even though fire service hose is flexible and therefore able to absorb much of the kinetic energy of the water, damage to couplings through too rapid shut-down of branches is possible and there is evidence that damage may also occur to pumps and collecting heads.

One particular activity where damage can occur is through the very rapid operation of the on/off control of high pressure hosereel diffuser branches, for instance, whilst adopting the "pulsing" technique as a firefighting tactic. This technique is designed to deliver a controlled amount of water, in droplet form, to cool hot gases. A slower, measured, operation of the on/off control to create the desired effect is preferred, so unnecessary damage to couplings etc. through water hammer should be reduced. (More information about using pulsing as a firefighting tactic can be found on page 14 of the Fire Service Manual, Volume 2 'Compartment Fires and Tactical Ventilation'.)

Some pumps have a small pressure relief channel in the non-return valve on the suction side of the pump which is designed to give protection to the collecting head against water hammer.

(ii) the rapid closure of the hydrant to tank valve or the hydrant valve itself

This may cause the main on which the hydrant is situated to fracture. Damage is most likely to occur when the main is of small diameter with a consequent high velocity of flow.

These examples indicate the necessity for slowly closing hydrants, shut-off type branches and other valves in order to avoid water hammer which might burst hose and damage couplings, pumps, collecting heads, tanks and water mains.

Chapter 2 – Instruments

Introduction

An understanding of the measurement of pressure is of obvious and fundamental importance in fire service work because the correct and safe use of pumps and comprehension of information on their performance and on the potential capacity of hydrants, etc., depend upon it. Sometimes, for example in the case of pressurisation systems on escape staircases, measurement of small pressure differences is important.

Flowrate measurements are another important aspect of hydrant testing. Also, in recent years, a number of firefighting appliances have been fitted with flowmeters and, although they are not in widespread use at present, it is appropriate that a limited amount of information on these is included. The advantages of the use of flowmeters in fireground pumping operations are discussed in paragraph 6.4.7.

2.1 Measurement of Pressure

2.1.1 Water gauges (manometers)

Very small pressure differences, such as those existing between compartments in buildings equipped with pressurisation systems, may conveniently and accurately be measured by means of a simple water gauge. This device (Figure 2.1) consists of a U-tube, one end of which is open and situated in one of the compartments whilst the other is connected, via flexible tubing to the other compartment. The water level in the open limb rises or falls depending on whether the pressure there is below or above the pressure in the closed limb and the vertical difference between the levels in the two limbs, measured in mm, is a convenient way of expressing the pressure difference.

The device may be refined in a number of ways aimed at increasing the accuracy with which it can

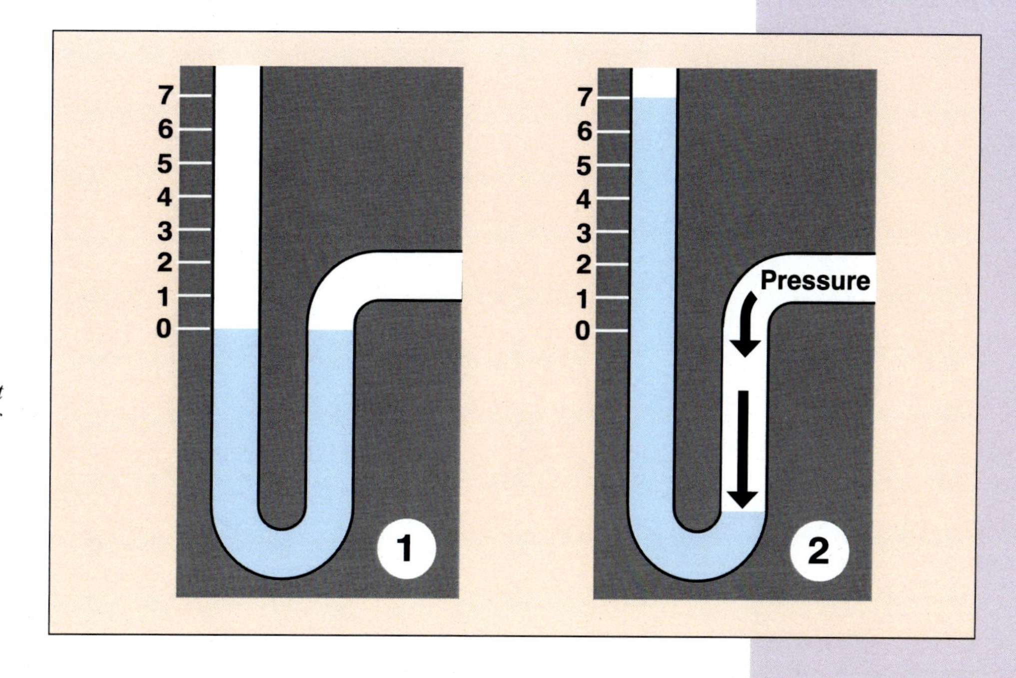

Figure 2.1 Diagrammatic arrangement of a water gauge showing: (1) reading when pressure difference is zero; (2) reading when pressure difference is applied. The scale is calibrated to indicate the difference in height between the surface of the water in the two limbs.

be read. For example, the graduated limb may be inclined rather than vertical resulting in the water surface having to travel much further along that limb in order to achieve the same vertical difference and so magnifying the separation between the graduations.

Liquids other than water, e.g. mercury, may be used depending on the range of pressure differences which it is required to measure.

For water, since the relationship between pressure and head is:

$$P = \frac{H}{10}$$

a difference in level of 1 metre would mean a pressure differential of 0.1 bar so:

each millimetre equates to a pressure differential of 0.0001 bar or 0.1 mbar.

2.1.2 Pressure and compound gauges

Most gauges used by the fire service indicate "gauge pressure" i.e. the difference between the true (absolute) pressure being measured and atmospheric pressure. They therefore register zero when exposed to atmospheric pressure but will give a positive indication when exposed to pressures above atmospheric and a negative indication when exposed to pressures below it.

Gauges may be constructed to measure pressure only, vacuum only, or may be arranged to read both. The first is termed a pressure gauge, the second a vacuum gauge, and the third a compound gauge. The water pumping system on fire service appliances is normally fitted with both pressure and compound gauges, whilst the engine lubricating system is equipped with a pressure gauge. The suction foam inductor is fitted with a vacuum gauge. There are two main types of gauge in use in the fire service, the Bourdon tube type and the diaphragm type.

(i) Bourdon tube gauge

In this type of gauge (Figure 2.2) the pressure-responsive element consists of a tube (the Bourdon tube), oval in section, sealed at one end and formed in the shape of the greater part of the circumference of a circle. When pressure is applied internally, this tube tends to straighten, and conversely when air is exhausted it tends to curl up more tightly. The amount of movement at the free end of the tube is approximately proportional to the pressure or vacuum applied and, provided that movement is relatively small, the tube returns to its normal position when the pressure or vacuum is released. A linkage at the free (sealed) end of the tube magnifies the movement and transmits it to a pointer moving over a suitably graduated dial. Gauges of this type used on fire brigade appliances conform to British Standard 1780. Their Bourdon tubes are usually made of phosphor bronze, the thickness of

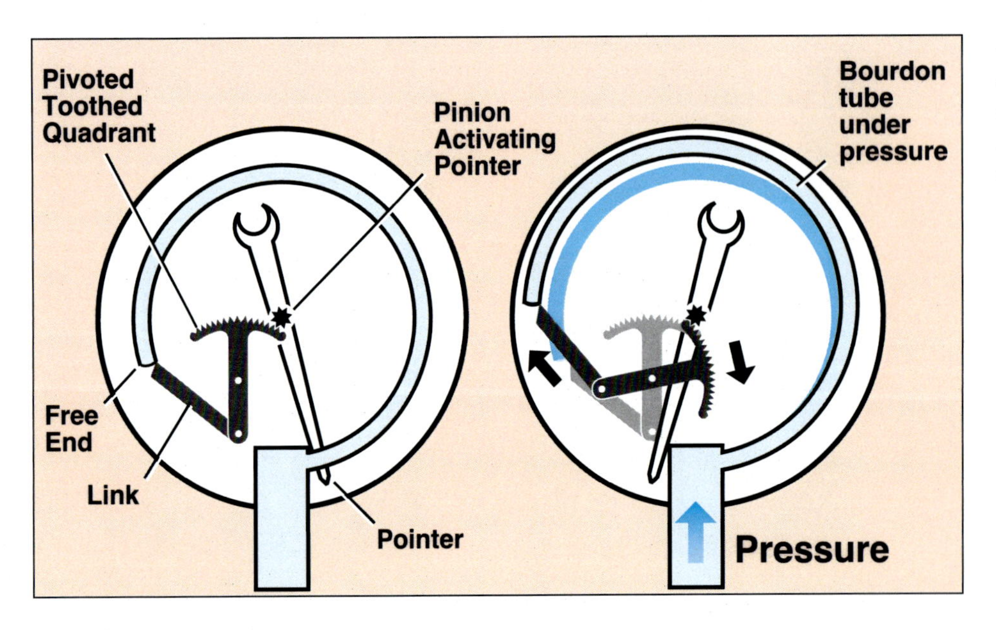

Figure 2.2 Diagrams showing (from behind) the mechanism of a Bourdon tube pressure or vacuum gauge.

the wall and the exact shape of the cross-section of the tube being carefully chosen for the pressure range over which the gauge is designed to operate. It is obvious that under no circumstances must the tube be subjected to excessive pressures, otherwise it would take on a permanent 'set' and thus become inaccurate.

The mechanism which connects the free end of the Bourdon tube to the pointer is called the movement. It usually consists of a pivoted toothed quadrant, one end of which is connected by a link to the end of the tube, whilst the teeth mesh with a pinion on the pointer spindle. Attached to the spindle is a hair-spring, the function of which is to keep the teeth of the pinion in close contact with those of the quadrant and so eliminate any free movement. The linkage can be arranged to magnify the scale reading as required. The geared movement allows an almost full circular reading to be obtained at the maximum pressure which the particular instrument is designed to measure.

The pointer of a bourdon gauge will move in a clockwise direction the same distance for a positive pressure of 1 bar as it will move in an anticlockwise direction for a vacuum of 1 bar so, only if the pressure scale is graduated up to just 1 bar, will both positive and negative scales on the gauge occupy the same length on the dial (Figure 2.3, left).

If, as is much more likely on fire pumps, the instrument is required to measure pressures up to 10 bar or more, the vacuum scale, reading down to -1 bar (i.e. absolute vacuum), would only occupy one-tenth or less of the length of the pressure scale (Figure 2.3, right) and would be difficult to read accurately. To overcome this difficulty the diaphragm gauge was introduced.

(ii) Diaphragm type gauge

This type of gauge depends for its operating principle on the movement of a corrugated metal circular diaphragm (Figure 2.4), which is secured at its periphery but free to move at its centre.

When subjected to a positive pressure the diaphragm may move only over a limited proportion of its surface into the small diameter cavity in the front housing, but, under reduced pressure, it is free to move over its whole surface into the large diameter cavity in the rear housing. The effect of this arrangement is that the diaphragm behaves much more flexibly under reduced pressure than it does under positive pressure and the resulting movement is correspondingly greater. This movement is transmitted via a peg which is attached to the diaphragm, a rocker bar, a quadrant and pinion to a pointer which moves over a graduated scale. The stabiliser and hair spring ensure there is no back-lash in the mechanism. Compound gauges of this type are specially designed to have a long vacuum scale and a relatively short pressure scale (Figure 2.5) and are used on the inlet side of fire pumps where it is

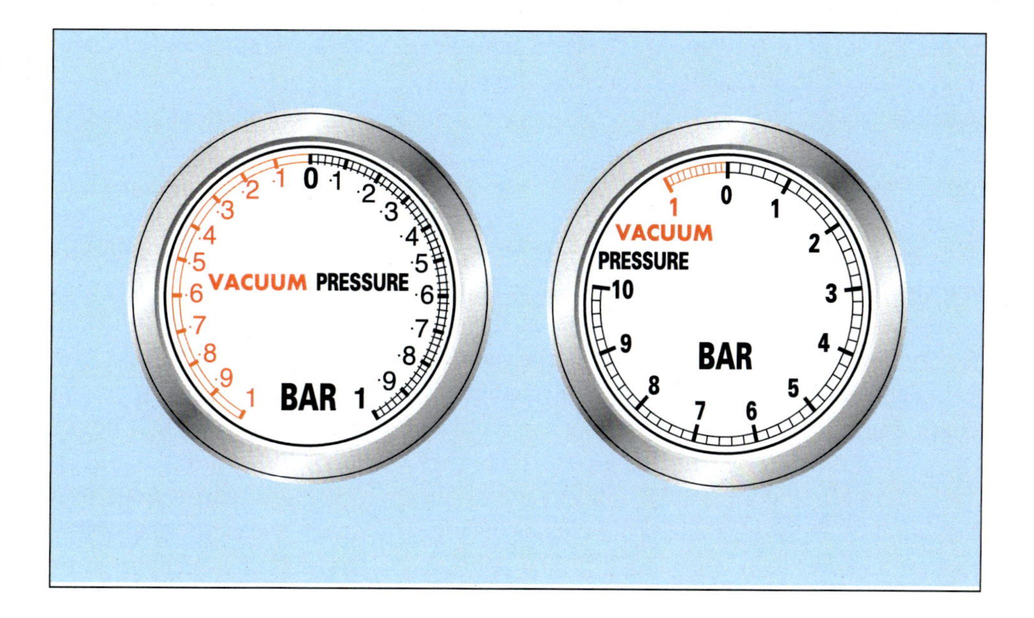

Figure 2.3 Bourdon tube compound gauge dials.

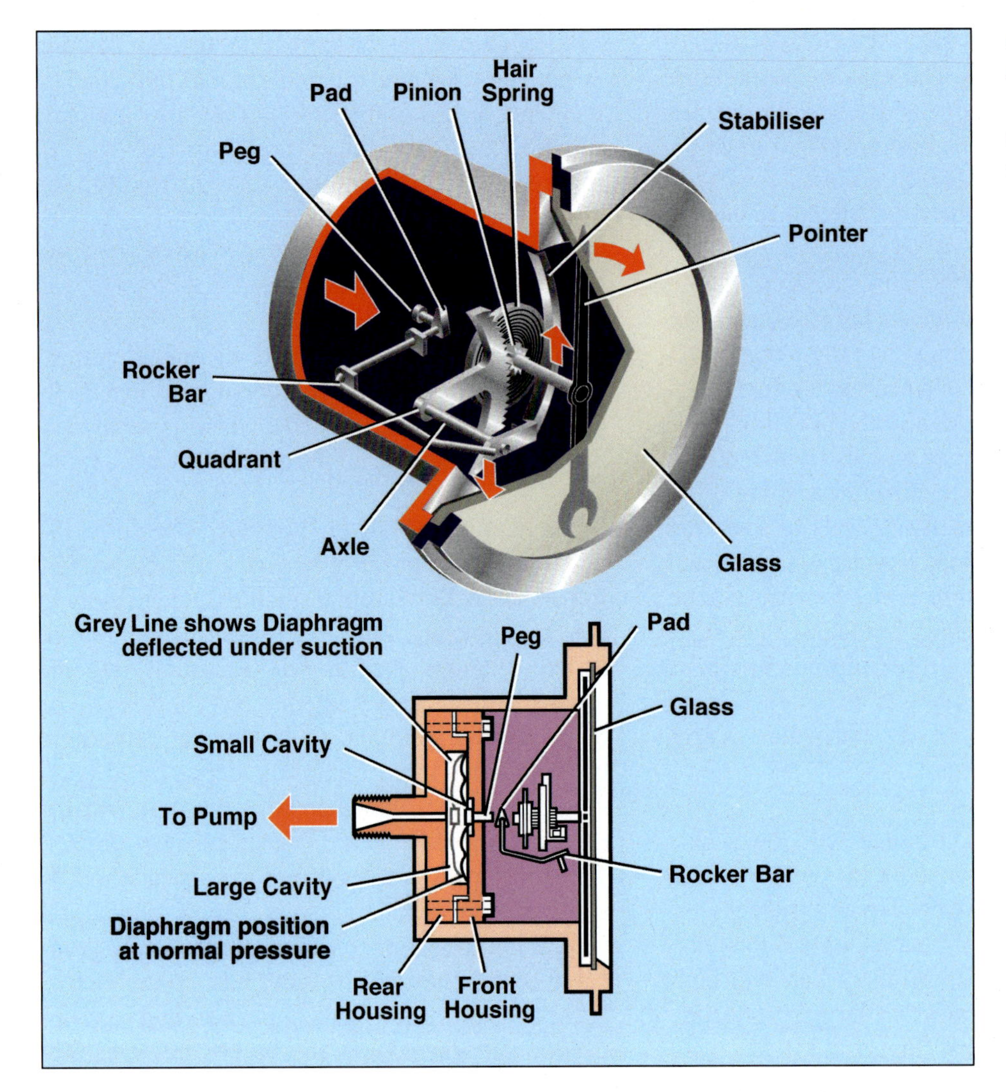

Figure 2.4 Diagram showing the construction of a diaphragm type compound gauge.

important to have a reasonably accurate measure of the incoming water pressure (which will be negative when pumping from an open supply below the level of the pump). There is, however, no relevant British Standard covering their construction.

(iii) Connections

Gauges on pumps are usually connected by means of copper piping to the point at which it is desired to measure the pressure, or the vacuum as the case may be. This piping usually has in it one or more coils designed to accommodate small changes in length due to vibration, expansion or contraction. These coils should always be arranged so that they are horizontal (Figure 2.6(1)). This prevents the formation of an air lock and water cannot become trapped and freeze (Figure 2.6(2)), thus splitting the pipe or at least causing a blockage and rendering the gauge inoperative.

(iv) Gauge cocks

The connections of both pressure and compound gauges are sometimes fitted with a cock at the point where they meet the pump casing. The functions of this cock are twofold: first, so that the pump can still be used if the gauge has been removed or the connecting piping has been damaged; and, second, to damp down the movement of the needle, where this is oscillating due to pulsation. It should be appreciated that manipulation of the gauge cock will not cure an unsteady reading due to vibration of the gauge itself. Where the reading fluctuates due to pulsation in the pressure to be measured, partially closing down the cock restricts the opening and retards the transmission

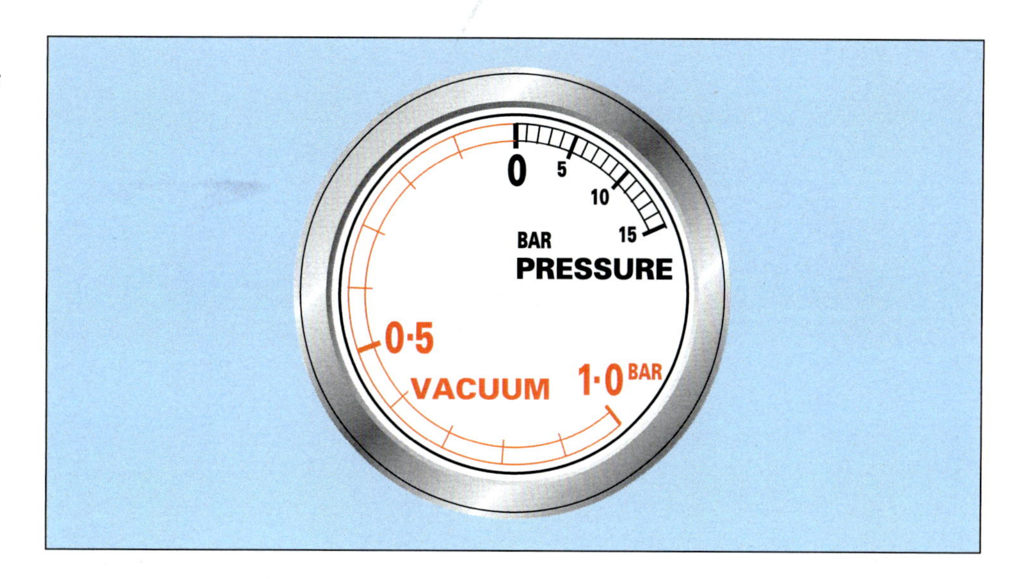

Figure 2.5
Diaphragm type gauge dial showing the long vacuum scale and relatively short pressure scale.

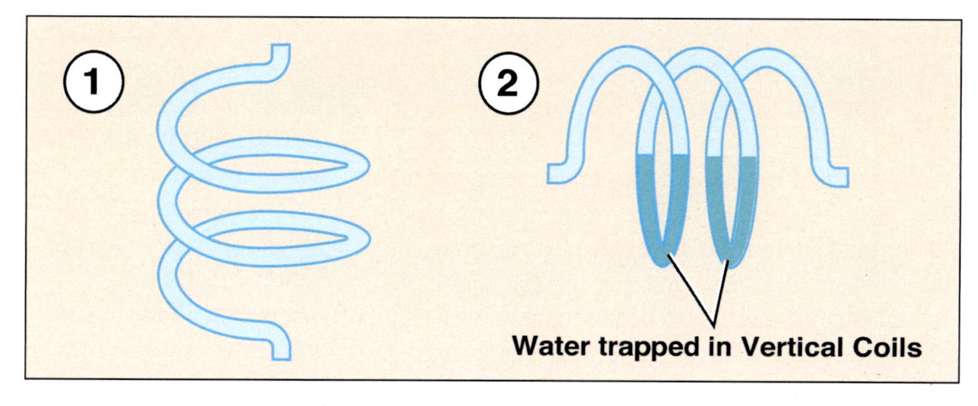

Figure 2.6
Diagrams showing:
(1) the correct arrangement for the coils on a gauge connection;
(2) incorrect arrangement leading to the formation of pockets of water etc.

of minor pressure fluctuations through the connection. The gauge thus shows a mean reading and follows the pressure variations more slowly than would be the case if the aperture were fully open. Under no circumstances should the cock be entirely closed in an endeavour to 'iron out' these fluctuations. This merely isolates the gauge from the pressure or vacuum which it is desired to measure and a false reading is obtained.

If, prompted by a low or zero reading on a gauge which has mistakenly been isolated, the pump operator increases pump pressure, a dangerous situation for branchholders may well result.

Other ways of dealing with the problem of rapid pressure fluctuations are:

(1) the inclusion of a throttling device, such as a fine capillary tube, in the gauge connection.

(2) filling the gauge with a viscous liquid, as shown in Figure 2.7, has the effect of damping fluctuations.

With the inclusion of one or other of these devices it is now quite usual to dispense with the gauge cock.

(v) Care of gauges

It should be appreciated that gauges are sensitive instruments and must be treated with care if they are to give long and accurate service without maintenance. Though failure of the pressure element is unlikely, shock resulting from sudden changes of pressure, should be avoided.

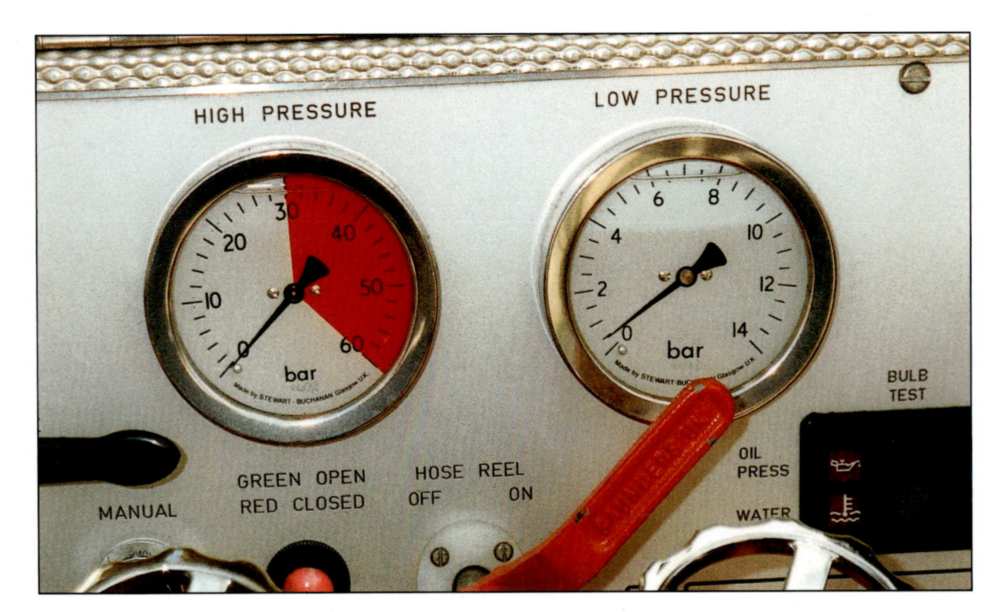

Figure 2.7 Viscous liquid damping of a pressure gauge. (Courtesy of Fire Service College)

2.2 Measurement of Flowrate

It is sometimes necessary to be able to measure the flowrate (the number of litres per minute or per second) available from a pump or hydrant and, for this reason, a limited number of the available instruments are described briefly below. Probably the most widely used of these instruments is the Vernon Morris hydrant testing gauge, but, in recent years, flowmeters have also been built into the deliveries of pumping appliances.

2.2.1 The Vernon Morris hydrant testing gauge

The principle of operation of the instrument, which incorporates a device called a Pitot tube, may be explained by reference to Figure 2.8.

The Pitot tube is a very simple device which is used to measure the velocity of flow of a stream of liquid. It consists only of a small diameter tube with a right angle bend in it fixed into the pipe which contains the flowing liquid as shown in Figure 2.8. The short limb of the tube is horizontal and points along the axis of the pipe towards the oncoming liquid whilst the longer limb is vertical and effectively forms a manometer.

As the moving liquid attempts to flow into the Pitot tube it forces its way up the vertical limb to a height which increases as the velocity increases. Although the relationship between the height to which the liquid rises in the manometer tube and the flowrate involves a number of factors, including the flow pipe diameter, it is quite easy to calibrate the manometer tube to read flowrate directly

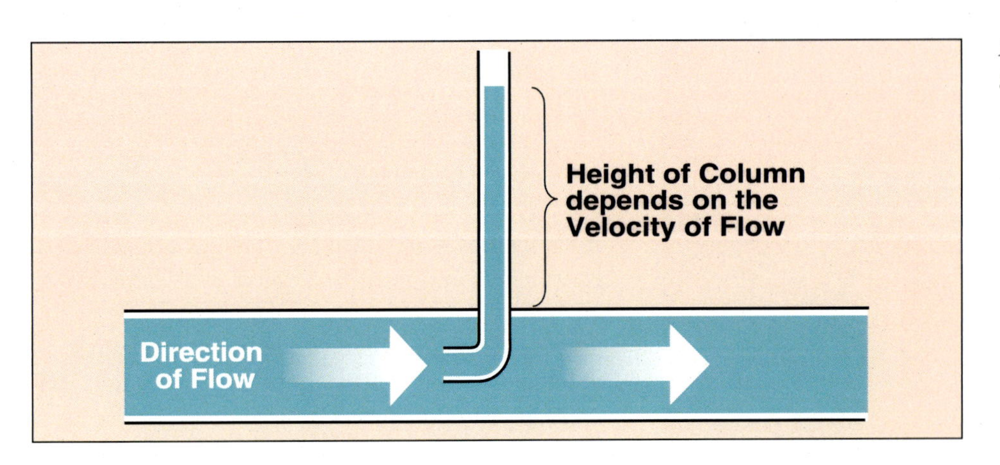

Figure 2.8 The Pitot tube principle.

for a standard size flow tube. However, to measure high flowrates, the manometer tube will need to be inconveniently long unless an alternative strategy, such as using a closed tube, is employed.

The Vernon Morris instrument, intended for hydrant testing, is shown in Figure 2.9.

It consists of a flow tube having a male connection at one end for attaching to a standpipe head, and a swivel joint at the other end to which the manometer is connected. The manometer has three ranges of flow marked on it in red, black and white: the standard range of 7.2 to 34 litres per second (l/sec) (432 to 2040 l/min); a low range of 2.6 to 7.2 l/sec (156 to 432 l/min), and a top range of 34 to 60 l/sec (2040 to 3600 l/min).

A pressure gauge graduated in metres head is also provided; this is attached to a male coupling so that the gauge can be connected to a standpipe head in order that the static pressure in the main can be checked. A vent cock is provided in the coupling to release the water pressure when the gauge is disconnected.

The following gives a brief outline of the operating principle and procedure:

Standard range (7.2 to 34 litres per second)

The vent plug is screwed tightly shut before flow commences. Because the air trapped in the manometer tube is now being compressed as the water attempts to rise, the range of the instrument

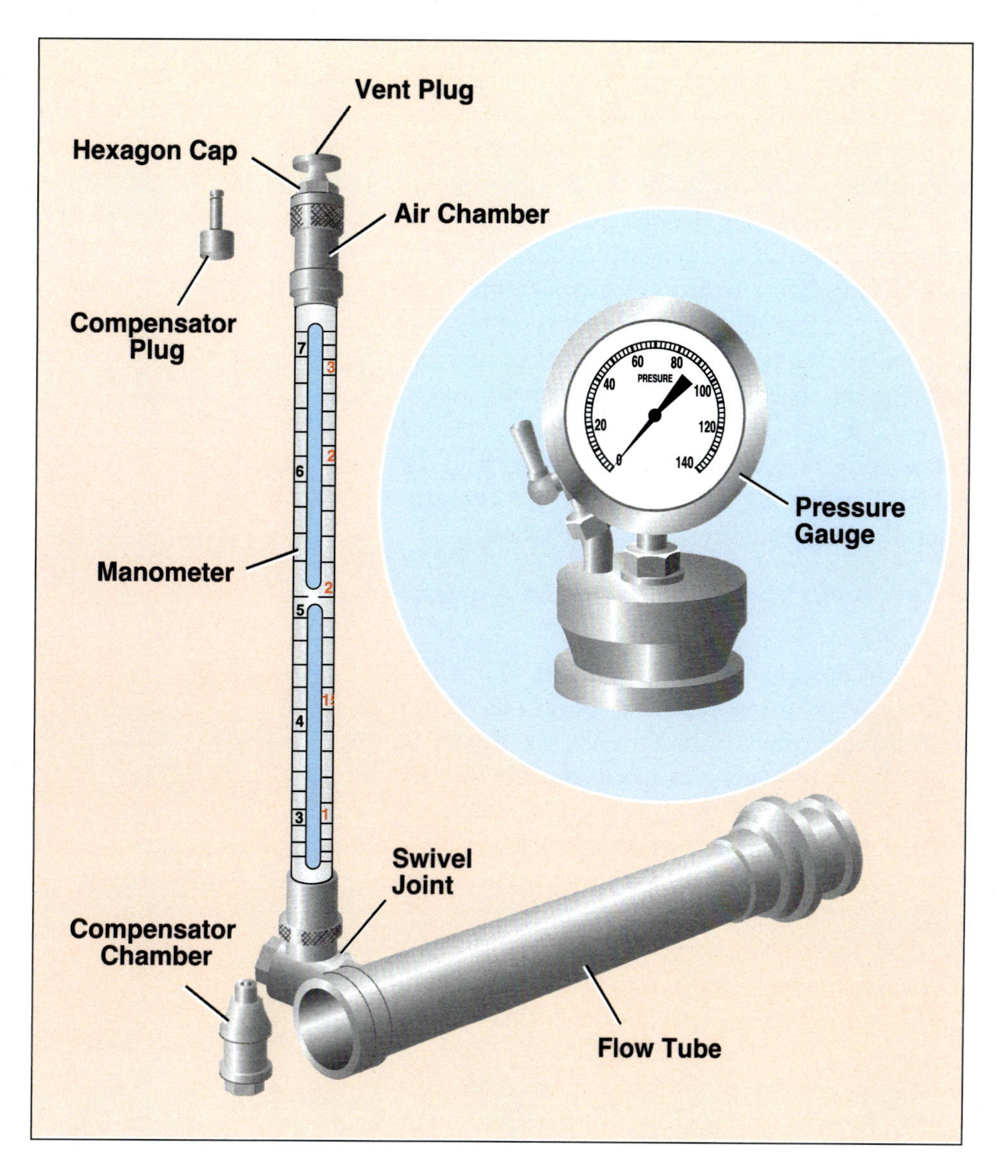

Figure 2.9 The Vernon Morris Flow Gauge for testing hydrants.

is effectively extended beyond that achieved with the tube open. The flowrate reading is obtained from the red scale on the manometer housing but, if the water level does not rise far enough to be visible, the vent plug is removed and the reading taken from the low range scale.

Low range (2.6 to 7.2 litres per second)

With the vent plug removed the manometer tube is open to atmosphere and flowrate readings are obtained from the black scale.

Top range (34 to 60 litres per second)

If, when using the standard range, the water level in the manometer tube goes off scale, then the flowrate exceeds 34 l/sec and the effective range of the instrument has to be further increased. In order to achieve this the hexagon cap is removed from the top of the manometer and the compensator plug inserted into the air chamber; the cap is then replaced and screwed tightly down. Before the manometer is attached to the swivel joint connection, the compensator chamber has to be screwed into the bottom of the manometer, and the combined fitting then attached to the swivel joint connection. This has the effect of increasing the volume of air trapped in the manometer tube but reducing the space into which it is compressed as the water level rises. The reading of the rate of flow is taken from the white figures on the back of the manometer housing. After use, both the compensator plug and chamber should be removed.

2.2.2 The paddle wheel type flowmeter

The instrument consists of a flow tube into which a paddle wheel is inserted so that the blades project partially into the stream of water as shown in Figure 2.10. It rotates at a speed which is directly proportional to the rate of flow of the water and creates magnetic pulses, which are counted electronically, each time one of the blades passes a sensor.

The instrument is calibrated to give a direct indication of flowrate and is sometimes fitted to the individual deliveries on firefighting pumps as shown in Figure 2.14. Figure 2.11 shows the Vernon Morris version of the instrument which is designed as an alternative to the hydrant testing gauge described previously. It has a useful range of 150 to 1800 l/min.

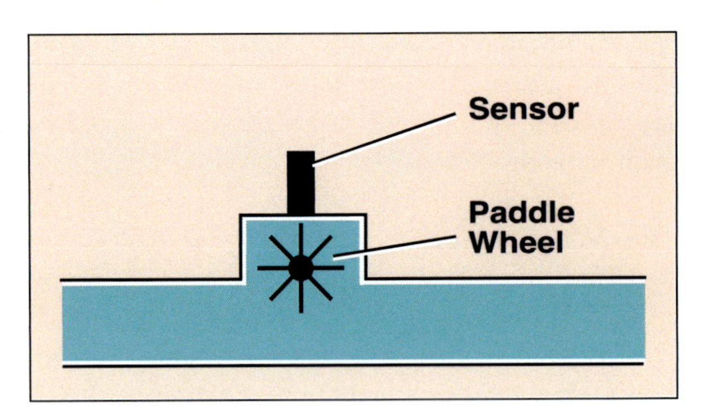

Figure 2.10 Operating principle of the paddle wheel type flowmeter.

2.2.3 Other types of flowmeter

A number of other types of instrument, of which just two are described below, are available for the measurement of flowrates from hydrants and from pumps. Although they are not widely used in brigades at the present time, their future use cannot be ruled out, so a brief outline of their operating principles is appropriate.

(a) Electromagnetic type flowmeters

When a conductor of electricity moves through a magnetic field which is at right angles to the direction of motion a voltage, proportional to the velocity of the conductor, is developed across it as shown in Figure 2.12. Water, even when

Figure 2.11 The paddle wheel type hydrant testing gauge in use. (Photo: Vernon Morris & Co. Ltd)

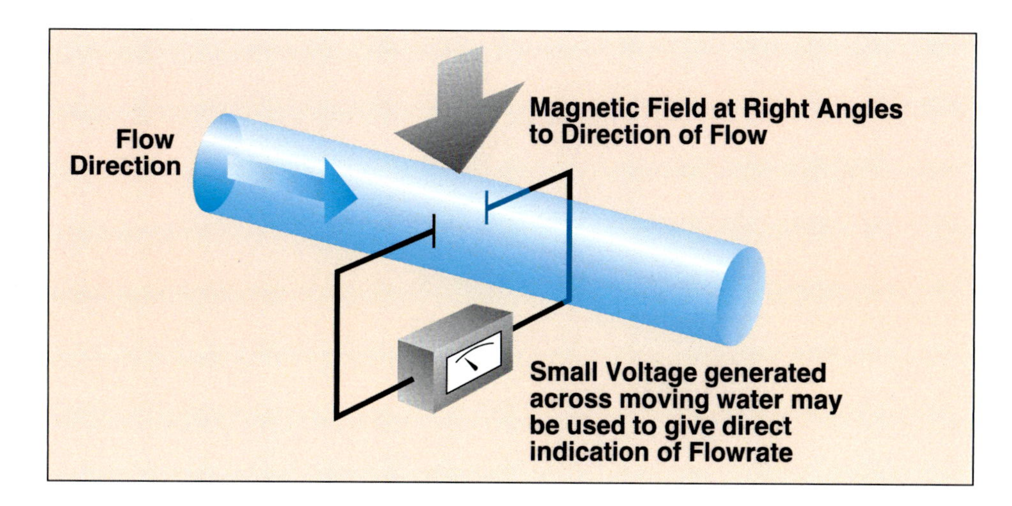

Figure 2.12 Principle of the electromagnetic flowmeter.

reasonably pure, is a sufficiently good conductor to cause a small voltage, which may be measured by a sensitive detector, to be generated and the flowrate, for a pipe of given diameter, may be indicated directly.

This type of instrument also is in use as a flow measuring device on some pumping appliances. One of its advantages is that there is no discernible pressure loss as water flows through it.

(b) Venturi type flowmeters

A typical venturi is shown in Figure 2.13. It consists of a section of a pipe in which there is a gradual smooth reduction in diameter, to a point called the throat, followed by an increase back to the normal diameter. As water, or any other fluid, flows through the device its speed increases to a maximum at the throat and then decreases to its initial value when the normal diameter is resumed.

As explained in paragraph 1.5.2, because of the increase in kinetic energy at the throat, the pressure P_2 at that point is less than the pressure P_1 at the point just before the diameter starts to reduce. The pressure difference between the two points, which may be measured by a suitable gauge, depends on the flowrate and can be used to give a direct indication of it.

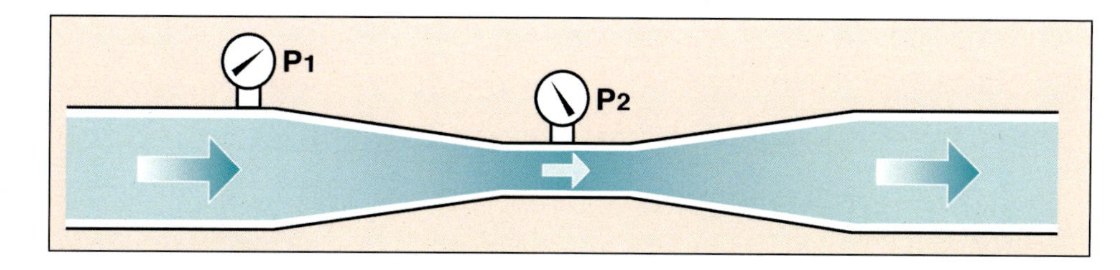

Figure 2.13 The flow of water through a venturi.

Figure 2.14 Flowmeters on a pumping appliance. *(Courtesy of Fire Service College)*

Hydraulics, Pumps and Water Supplies

Chapter 3

Chapter 3 – Atmospheric Pressure and Suction Lift

Introduction

The atmosphere which envelops the earth, although gaseous, possesses definite weight. Since the weight is exerted, for all practical purposes, uniformly all over the surface of the earth and through the cavities of the body as well as externally, the human body is unaware of it and behaves as though such pressure did not exist.

However, it is important to understand the significance of atmospheric pressure because the height to which a pump can lift water under suction from an open source, the performance of the pump and the operation of a siphon are all dependent on it.

3.1 Atmospheric Pressure

A column of air 1 metre square and extending to the upper limit of the atmosphere weighs, on average, 101 300 newtons at sea level, and, therefore:

Atmospheric Pressure = 101 300 N/m² or 1.013 bar

(remembering we must divide by 100 000 to convert from N/m² to bars)

For most practical purposes, therefore, we may regard atmospheric pressure as having a value of 1 bar

Atmospheric pressure acts on the surface of any open body of water just as on the other parts of the earth's surface. If, however, this atmospheric pressure is removed from above a portion of the water, in other words, if a vacuum is formed, then the surface of the water is subjected to unequal pressures and, the water being resistant to compression but not to change of shape, that part of the surface which is relatively free from pressure will tend to rise. The behaviour of water in this way, when subjected to a partial vacuum, is of fundamental importance in firefighting operations, since upon it depends the process of priming a pump from open water.

It will be convenient at this stage to study Figure 3.1. In the diagram AB is the surface of a sheet of open water over part of which, CD, is inverted a long tube, sealed at the upper end, except for a connection to a pump. If, initially, the pressure in the tube acting on the surface CD is normal atmospheric the system may be regarded as akin to a pair of scales, the two pans of which are equally weighted, i.e. each is loaded with 1.013 bar. If the pressure in the vertical tube is now reduced, the pressure bearing down on the portion CD is progressively

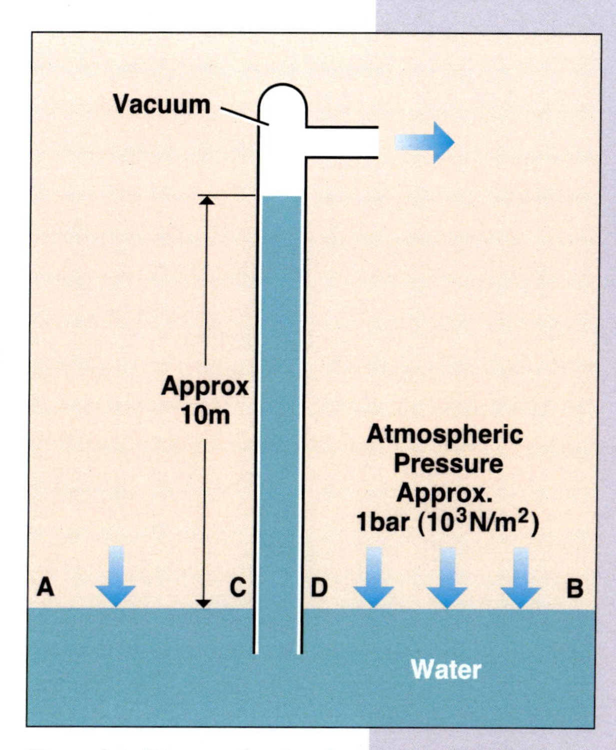

Figure 3.1 Diagram showing the principle of lifting water by atmospheric pressure.

reduced in value. The atmospheric pressure outside the tube does not change, and it will be seen, therefore, that the system is now unbalanced. Water flows up the tube until the pressure exerted at the base of the column of water in the tube is equal to the pressure of the surrounding atmosphere.

Theoretically if an absolute vacuum could be formed in the tube, i.e. if the air could be completely exhausted, then the water would rise a vertical distance sufficient to create a pressure of 1.013 bars at the level CD.

Using the relationship, already established, between head and pressure:

$$H = 10.19 \times P$$

this vertical distance is seen to be:

$$10.19 \times 1.013 \text{ or } 10.3\text{m}$$

No amount of further effort, whether by endeavouring to increase the vacuum (which is impossible) or by lengthening the tube, will induce the water to rise higher than 10.3m. It has already been stated that sea water is slightly denser than fresh water, so that a shorter column of sea water would balance the atmospheric pressure. Because the density of sea water is 1.03 kg/litre, it would require a column of length:

$$\frac{10.3}{1.03}$$

i.e. 10 metres to balance the atmospheric pressure.

If, however, the air is not completely exhausted from the tube, then the water will only rise to a height sufficient to balance the pressure difference existing between the inside and the outside of the tube. Thus, if there remains in the tube air at a pressure equivalent to that exerted by a column of water 3.3m high, then the water will only rise 7m.

It will be seen therefore that when a vacuum is formed in a long tube or pipe, water is driven up by the external pressure of the air acting on the exposed surface of the water although, in common parlance, we say that it is sucked up by the vacuum.

This is precisely what happens when a pump is primed. The suction hose takes the place of the vertical tube and the primer is the device for exhausting or removing the air from the suction. As the pressure inside the suction is reduced, so the excess pressure of the air on the exposed surface of the water forces the latter up the suction until it reaches the inlet of the pump. It follows, therefore, that even theoretically, when under perfect working conditions, at normal atmospheric pressure a pump cannot lift water from a greater depth than 10.3m from its surface to the centre of the pump inlet.

The value for atmospheric pressure quoted above, i.e. 1.013 bar, is the average value at sea level. It changes, according to variations in meteorological conditions, by as much as about 5% below or above this figure, so the maximum suction lift achievable when a region of high pressure is over the country may be almost a metre more than when there is a deep depression. There is also a significant variation of atmospheric pressure with altitude. It reduces by approximately 10% (0.1 bar) for every 1000m gain in height so that maximum theoretical suction lift is correspondingly reduced – to about 9m at an altitude of 1000m.

It has already been shown that water will rise to a height of 10.3m in a completely evacuated tube, and such a device could, in principle, be used as a barometer to measure atmospheric pressure. However the height of the instrument would make it entirely unpractical for general use, so the length of the column necessary to counterbalance the air pressure is reduced by using the densest convenient liquid – mercury. As mercury is 13.6 times as dense as water, atmospheric pressure will be capable of supporting a column of height 10.3/13.6 metres, i.e.

atmospheric pressure = 0.76m or 760mm of mercury

3.2 Suction Lift

3.2.1 Basic facts

A pump is said to lift water when the water is taken from an open source below the inlet of the pump and the suction lift is the vertical distance between the water surface and the centre line of the impeller. Water has no tensile strength and cannot therefore be pulled upward, and as we have already seen, it is the atmospheric pressure only which raises the water (see Figure 3.1).

The purpose of a priming device, when water is being lifted on the suction side of a pump, is to create a partial vacuum within the pump chamber and suction hose. The atmosphere exerts pressure on the open surface of the water and so forces the water up through the suction hose and into the pump. The mechanical condition of the pump and hence its ability to create a partial vacuum also has a bearing on the total height of the suction lift.

3.2.2 Practical Limitations

It has been shown that water cannot rise to a vertical height greater than approximately 10m in a completely evacuated tube, and that it does so because it is forced up by the atmospheric pressure acting on the water surface outside.

When pumping from an open source it is the atmospheric pressure alone which can provide the lift necessary for the water to enter the pump.

However, the atmospheric pressure's ability to lift water is reduced to below the theoretical capability of about 10m because it also has to:

(i) Create flow i.e. to give the water kinetic energy as it changes from its static state in the open supply to its moving state in the suction hose;

(ii) overcome frictional Resistance to flow in the suction hose and couplings;

(iii) overcome pressure loss due to turbulence and shock as the water enters the pump impeller. It is known as Entry loss and varies with the design of the pump;

(iv) overcome pressure loss as the water is forced through the Strainer and changes its direction of flow after entering;

(v) overcome any tendency for the water to vaporise (and so create an opposing vapour pressure) as it nears the impeller. This tendency increases rapidly with Temperature. In fact the vapour pressure of water becomes equal to atmospheric pressure at its boiling point so making it totally impossible to lift water at, or near, this temperature. At normal outdoor water temperatures the effect is small, however.

These five factors may conveniently be remembered by the mnemonic **CREST**.

Factors (i), (ii), (iii) and (iv) are each increased by increasing flowrate, and because the proportion of atmospheric pressure available to deal with them decreases as lift increases, it follows that the output from the pump will also decrease with increasing lift. Whilst suction lifts of 8.5m or more are sometimes obtained under very good conditions, **8m** can be considered the approximate practical maximum if a worthwhile output is to be obtained from the pump.

In order to appreciate the impact of lift on pump performance consider a pump working from open water with a suction lift of 3m as shown in Figure 3.2 left. (Note that when manufacturers give information on pump performance, a subject discussed in Chapter 5, it is common to specify a lift of 3m.) To do the work of raising the water to the pump a pressure equivalent to 3m head of water is required. This, subtracted from a total available pressure equivalent to 10m, leaves a pressure equivalent to 7m head to overcome the various losses described above.

If, however, the suction lift were 8m (Figure 3.2 right), a pressure equivalent to 8m head of water would be required to raise the water to the pump, leaving only 2m head to overcome the losses. Consequently, for any given diameter of suction, the quantity of water capable of being delivered would be considerably less than that possible at a 3m suction lift.

The practical steps to be taken in order to ensure the best possible pump performance when pumping from an open supply are discussed in Chapter 6.

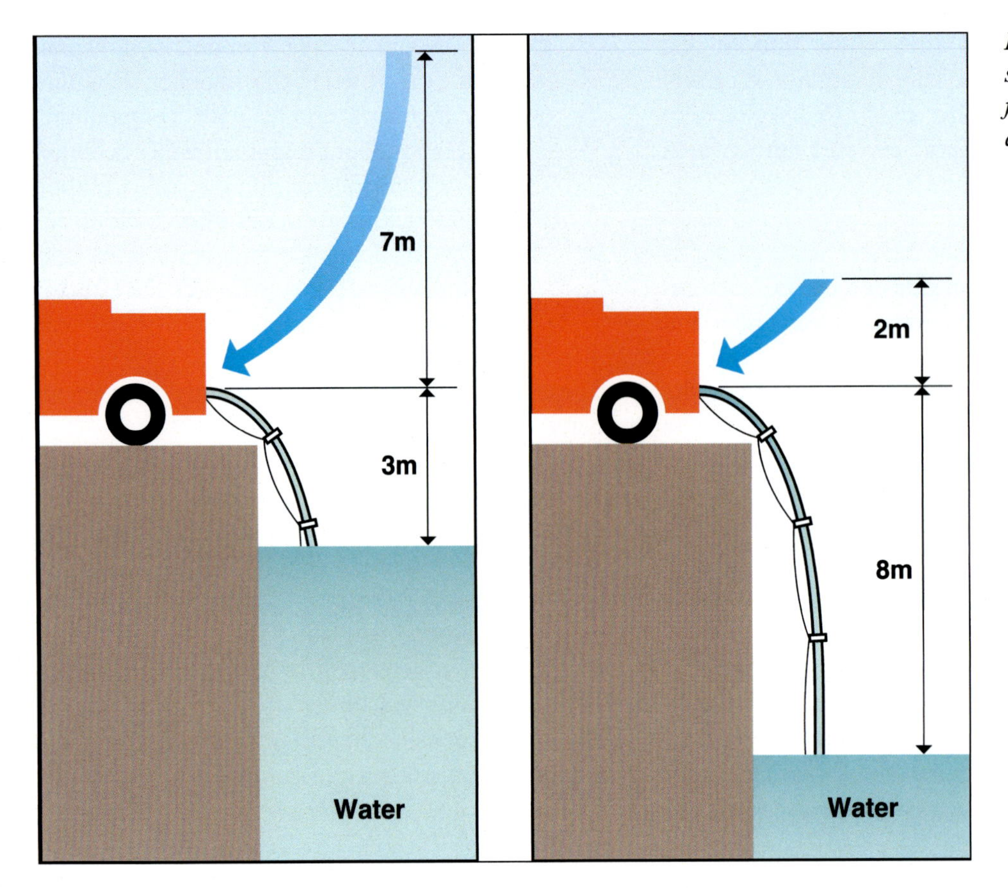

Figure 3.2 Sketch showing pump working from a lift of (left) 3m, and (right) 8m.

3.3 Siphons

A siphon (Figure 3.3) consists of an inverted U-tube with legs of unequal length and is used to transfer liquid, e.g. over the edge of the vessel in which it is contained, to a point at a lower level. One leg of the siphon must be submerged in the liquid which is to be lifted, and the point of discharge must lie at a lower level than the surface of the liquid. Its operation may be compared to the behaviour of a chain suspended over a pulley with unequal vertical lengths. The longer length, because it is heavier, will pull up the shorter length even if both ends are resting on horizontal surfaces. However, the important difference between a chain and a column of liquid is that the latter has no tensile strength and maintains its integrity only because of atmospheric pressure. It follows that, when using water, the siphon will only work so long as the shorter limb is less than 10m. The pipe or hose used must be rigid because otherwise, the pressure inside it being less than atmospheric, it will simply collapse under the atmospheric pressure acting on the outside.

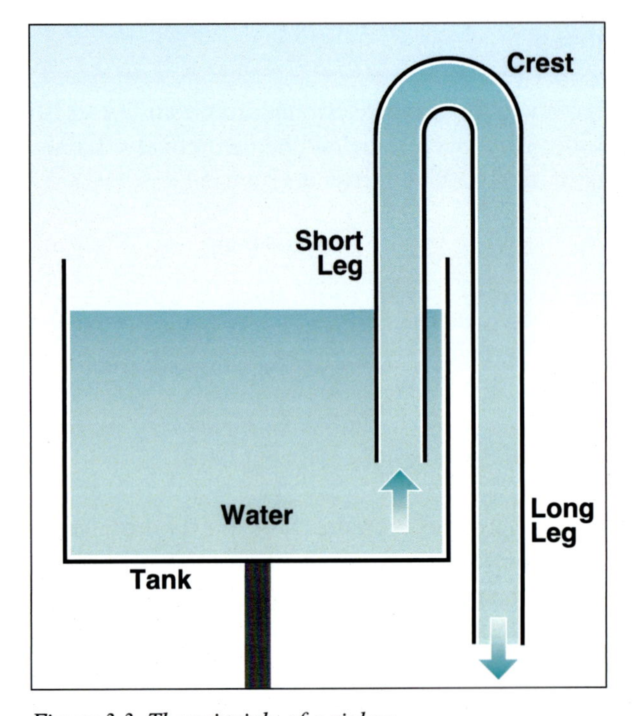

Figure 3.3 The principle of a siphon.

Chapter 4 – Water Supplies and Hydrants

Introduction

This chapter discusses legislation concerning water for firefighting, the water authorities' methods of water distribution and water supplies especially for firefighting purposes. This subject naturally leads on to hydrants, and related equipment, which give brigades the means to access these supplies.

4.1 Legislation concerning Mains Water Supplies

4.1.1 The Water Industry Act 1991, Water Act 1989, Water (Scotland) Act 1980, Northern Ireland

The Water Act 1989 (which applies only to England and Wales) provided for the former water authorities to be privatised and established the appointment of a Director General of Water Services to ensure that the water companies are able to finance their functions and to determine and monitor levels of customer service.

Independent of the Director, Regional OFWAT Customer Service Committees were also set up, comprised of appointed members, to promote the interests of customers.

The Act also set new standards for the supply and quality of water. In 1991 the clauses of the 1989 Act which related to the provision of water for firefighting were included in the Water Industry Act of that year. The statutory requirements of the 1991 Act will be discussed later in this chapter.

Water undertakers have a duty to provide a supply of water for 'domestic purposes' and an obligation to provide a supply of water for commercial or industrial purposes if it does not affect the domestic supply of existing customers.

There are now many performance indicators used to report upon undertakers together with mandatory targets for leakage and water quality compliance.

The Environment Agency has a water resource licensing authority in addition to pollution monitoring and control, which through abstraction licensing, can have an impact upon the amount of water available in distribution systems.

The Drinking Water Inspectorate has responsibility for monitoring water quality and instigates prosecutions in such events as the supply of discoloured water which is considered 'unwholesome'.

Currently, the Water (Scotland) Act 1980, as amended, is the primary legislation covering this subject in Scotland.

In Northern Ireland responsibility for water resources lies with the Department of the Enviroment (Northern Ireland).

4.1.2 The Fire Services Act 1947

Sections 13, 14, 15 and 16 of the Fire Services Act 1947 also contain provisions for the supply of water for firefighting.

Section 13 of the Act requires that:

> **"A fire authority shall take all reasonable measures for ensuring the provision of an adequate supply of water, and for securing that it will be available for use, in case of fire."**

Under Section 14 (1 and 2) of the Act, the fire authority may enter into an agreement with statutory water undertakers for the purpose of implementing Section 13 on such terms as to payment or otherwise as may be specified in the agreement. No water undertakers shall unreasonably refuse to enter into any such agreement proposed by a fire authority. Section 14 (3) of the Act states that undertakers are responsible, at the expense of the fire authority, for the clear indication, by a notice or distinguishing mark which may be placed on a wall or fence adjoining a street or public place, of the situation of every fire hydrant provided by them.

The obligations of statutory water undertakers under subsections (1), (2) and (3) are enforceable, under section 18 of the 1991 Water Industry Act. Any complaints about a water undertaker not carrying out these obligations should be directed to the Secretary of State if they cannot be resolved.

A fire authority is empowered under Section 15 (1) of the Act to make agreements to secure the use, in case of fire, of water under the control of any person *other than water undertakers*; to improve the access to any such water and to lay and maintain pipes and carry out other works in connection with the use of such water in case of fire. Section 15 (2), however, indicates that the fire authority may be liable to pay reasonable compensation for this use.

Section 16 of the Act provides that if a person is proposing to carry out works for the supply of water to any part of the area of a fire authority that person is required to give not less than six weeks notice in writing to the fire authority prior to the commencement of the works. In the case of works affecting any fire hydrant the authority or person executing the work is normally required to give the fire authority written notice at least seven days before the work has begun. In an emergency, where it would be impracticable for notice to be given in the time stipulated, notice is to be given as early as may be.

Section 30 of the Act establishes that, at any fire, the senior fire officer present has sole charge and control of all operations for the extinction of the fire, including the use of any water supply, and that the water undertaker, on being requested by this officer to provide a greater supply and pressure of water for firefighting, shall take all necessary steps to comply with the request even if this results in the interruption of supplies to normal consumers. Failure, without reasonable excuse, on the part of the undertaker to comply is an offence.

NB. Section 147 of the Water Industry Act 1991 stipulates that no charge may be made by any water undertaker in respect of

(a) water taken for the purpose of the extinguishing of fires or taken by a fire authority for any other emergency purposes;

(b) water taken for the purpose of testing apparatus installed or equipment used for extinguishing fires or for the purpose of training persons for fire-fighting; or

(c) the availability of water for any purpose mentioned in paragraph (a) or (b) above.

However, Section 23 of the Water (Scotland) Act 1980 states only that:

> "the undertakers shall allow any person to take, without payment, water for extinguishing fires from any pipe on which a hydrant is fixed."

4.2 Distribution of Water Supplies

4.2.1 General

Water undertakers obtain their water from three main sources:

(i) River intakes.
(ii) Impounding reservoirs. These contain water collected from high ground, streams and general rainfall.
(iii) Underground sources e.g. wells, boreholes and springs.

About one-third of the total supply is drawn from each source but in each case the water is fed into main storage reservoirs, purified and then passed into the distribution system. This system conveys water to the consumer and, in general, consists of mains and pipes laid under public roads. There is no standard pattern for an authority's distribution network but it will often consist of:

- **a network of trunk and distribution mains, service reservoirs and booster pumps (where necessary) which supply water to:**

- **a number of zones, sometimes referred to as District Metered Areas (DMAs), in each of which the pressure is maintained at a high enough value to satisfy normal consumer demand.**

Each DMA (Figure 4.1) is supplied by one or two incoming mains, sometimes with pressure reducing valves, and the entry point(s) metered so that the consumption of the DMA may be measured. If it is necessary to shut down the usual supply main for maintenance or other reasons, water may be imported from a neighbouring DMA via the zone valves. A sudden dramatic increase in consumption may well indicate that a burst has occurred, the location of which can be determined by closing down different sections of the system until a reduction to normal consumption resumes.

The design of the public network, including diameters of pipes etc., is based on factors such as:

- Maximum requirements at peak demand.
- Minimum pressure requirements.
- Ageing factors.
- Estimated future demands.

Figure 4.2 shows typical daily variations in demand for a week in the winter and for a week in the summer.

The DMAs are supplied by a system of trunk mains (anything up to 1.2m in diameter) and distribution mains. Service mains, which supply the consumers within a DMA, are smaller and

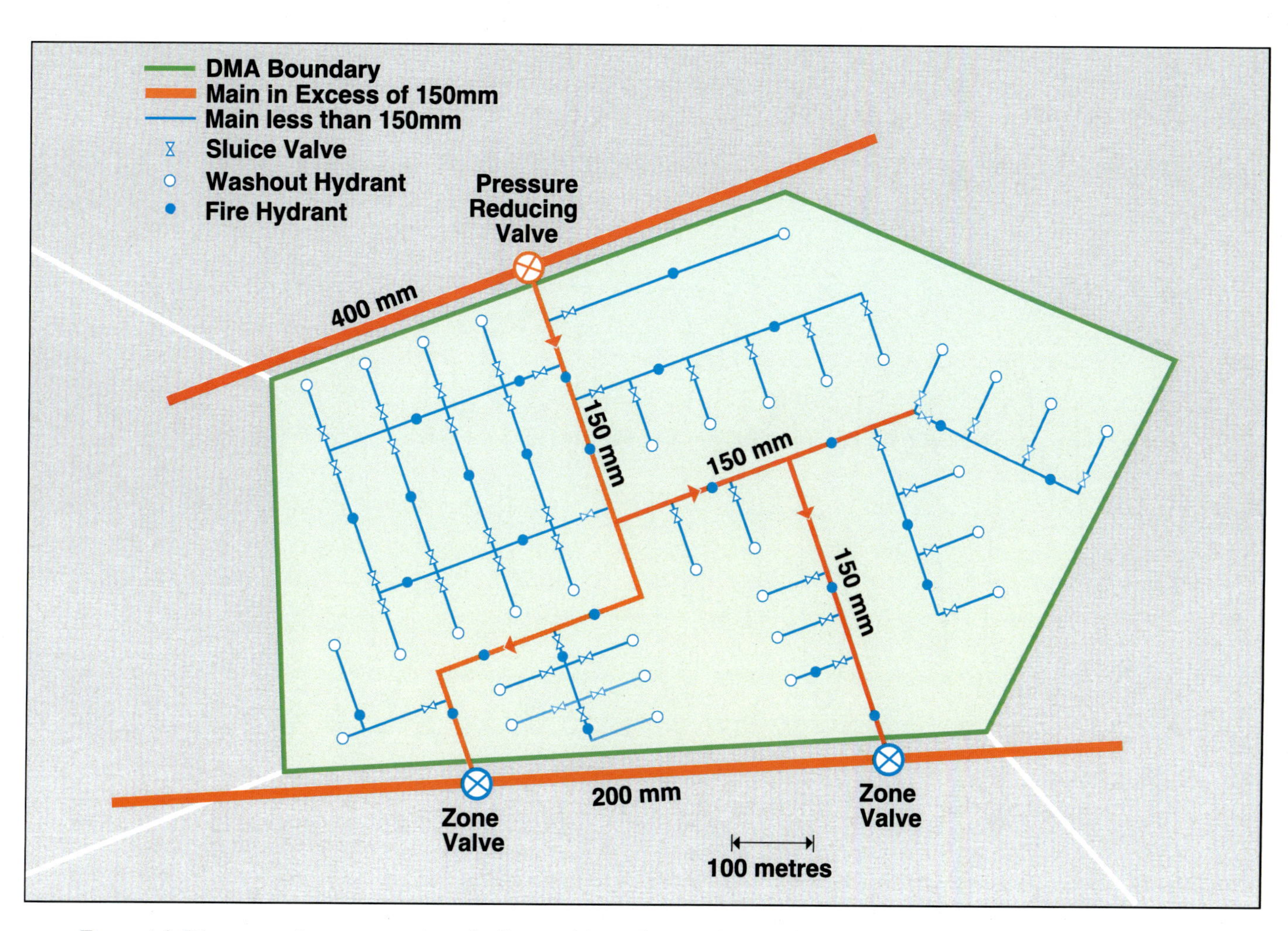

Figure 4.1 Diagrammatic representation of a District Metered Area. The mains are normally laid under public roads.

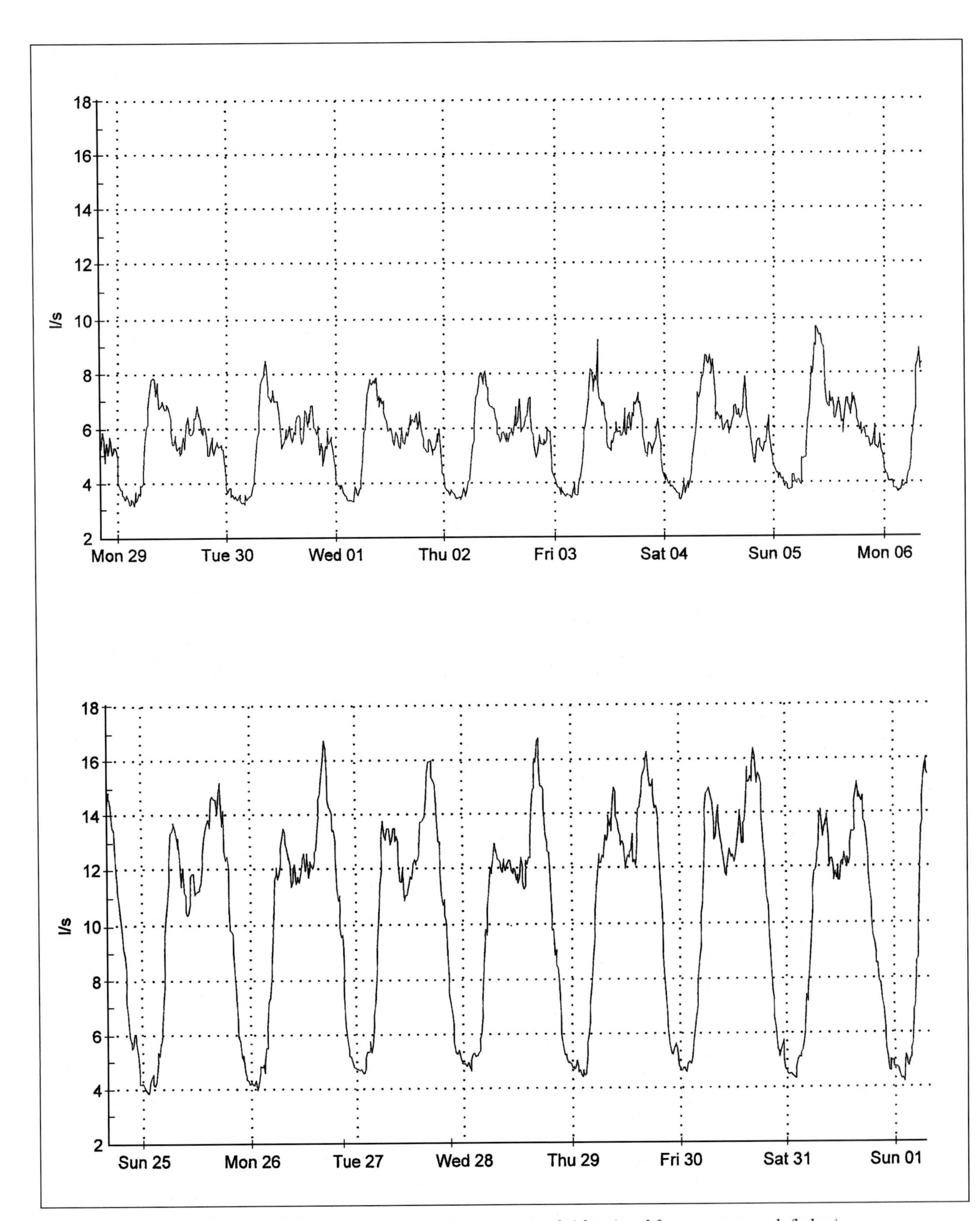

Figure 4.2 Typical daily variations in water demand for a winter week (above) and for a summer week (below).
(Courtesy of Wessex Water)

usually of 75, 100 or 150mm diameter. Large consumers, whose demands may be too great for the service mains, may sometimes be supplied by distribution mains. In rural areas even the distribution mains may not exceed 150mm and could be only 75mm.

4.2.2 Associated Features

(a) Sluice valves

Sluice valves are fixed at intervals along trunk and distribution mains, and occasionally on service mains if they are larger than usual. (Note: the term 'stop valve' is normally used for valves in domestic premises.)

Valves that are normally closed are used to separate areas with differing pressure characteristics, areas that fall into different water supply or water quality zones, or that define District Metered Areas. Most valves are normally open and are placed on the system to enable areas of the network to be isolated for repair, maintenance or emergency works.

They are operated by water undertakers' officials to control or isolate flows efficiently within the distribution system. **They can also be used, in some cases, to divert water where there is a shortage for firefighting.**

(b) Service reservoirs

Service reservoirs serve the dual purpose of balancing the distribution system and providing a reserve of water against the possibility of an interruption in supply due to a breakdown or excessive demand. These often include large overhead tanks and water towers.

(c) Booster pumps

Booster pumps are used to increase pressure in trunk mains for transfer of supplies over long distances.

And, in addition, to:

(i) provide opportunities to reconfigure areas to balance pressures;

(ii) provide support to areas, particularly at times of peak demand, when gravity flows would be insufficient to maintain supplies;

(iii) provide supplies to areas that lie close to, or above, the level that water is physically able to flow under gravity alone.

(d) Pressure reducing valves

Except for those in which the pressure is already low, because they lie close to reservoir level, typical DMAs would, in order to minimise the amount of water wasted through leakage, have pressure reducing valves on each of the incoming mains to lower the operating pressure to the minimum needed to sustain normal requirements. Though section 65(1) of the 1991 Act requires only that the pressure be sufficient to cause water to reach the top of the top-most storey of every building within the undertaker's area, OFWAT, the regulatory body for the water industry, sets the minimum pressure at the boundary as 10 metres head at a flow of 9 litres/minute.

'Pressure Management' is increasingly being employed using more sophisticated devices which maintain a minimum pressure but open up in response to increases in flow, thus maintaining target pressures.

4.2.3 Deterioration of Pipes

Cast iron or other ferrous metal pipes often develop tuberculation or corrosion internally, which increases the friction loss in the main and therefore reduces its carrying capacity (see Chapter 1). The design of the distribution system may have to take this into account, based on the calculated severity of the expected deterioration.

Plastic pipes, which are not susceptible to tuberculation and corrosion, are being increasingly installed as also are special internal linings, similar to fire hose, as a means of re-conditioning badly corroded and leaky mains.

Water undertakers maintain regular refurbishment programmes to overcome these problems and improve running pressures and the flow carrying capacities of their networks.

4.3 Water Supplies for Firefighting

4.3.1 General

Section 13 of the Fire Services Act specifies the fire authorities' responsibility for securing adequate supplies of water and ensuring that it is available for firefighting. In the United Kingdom most of this water is taken from water undertakers' mains though water from private mains, e.g. on factory premises, is also used, subject to agreement. However the primary function of these mains is to provide water for domestic purposes and this provision may not be adequate for firefighting in a particular area.

There is, at the present time, no legislation which requires water undertakers to provide minimum flowrates for firefighting purposes.

Mains supplies can be improved by increasing the size of the mains but, if this is requested by a fire authority, then that fire authority will have to pay. Financial considerations will often limit the sort of improvement that is reasonable for a fire authority to ask for, so it may be necessary to negotiate a compromise with the water undertaker between firefighting requirements and what the undertaker would provide for domestic purposes.

To augment mains supplies, brigades should survey all sources of water, including open water, near enough to to be of use for fire risks in their area (see Chapter 7 – Pre-planning). This has become even more important in recent years as water authorities have lowered operating pressures, particularly at night when ordinary domestic requirements are at a minimum, to reduce both the probability of leaks occurring and the high proportion of water which is wasted when they do.

Because of the possibility of contamination, water should never be taken directly from service reservoirs though it may be acceptable to the water undertaker for it to be taken from adjoining washouts.

All brigades keep records of the water supplies in their area and the access to them. As the Fire Services Act 1947, section 16, requires fire authorities to be notified of any intention to carry out works in connection with the supply of water, it should, therefore, be possible to keep these records up to date.

4.3.2 Agreements between Fire Authorities and Water Undertakers

Many fire authorities have negotiated codes of practice (based on a model agreement drafted in 1993 by the Chief and Assistant Chief Fire Officers' Association) with the relevant water undertakers to establish a working relationship between them and so to enable fire authorities to meet their obligations under Section 13 of the Fire Services Act. These codes of practice cover areas such as:

- Hydrant installation, maintenance and testing.
- Provision of water for firefighting purposes.
- Liaison between the fire authority and the water undertaker, particularly during emergencies.
- Charges to the fire authority for work on hydrant installations etc.

Problems regarding water for firefighting may still occur and, in order to help resolve them, a **National Liaison Group** was recently set up. This had the brief of drafting a new Code of Practice to "facilitate and promote liaison between Local Authority Fire Brigades and Water Companies by producing guidance which identifies the issues that Water Companies and Fire Brigades should consider when preparing their own local agreements".

The resulting "National guidance document on the provision of water for firefighting" was published jointly by the Local Government Association and Water UK in December 1998. It places particular emphasis on improved local, district and national liaison between the brigades and water undertakers and recommends that local discussions should address the following areas:

- Facilitating the ongoing availability of water for firefighting.

- Communication in respect of and during an emergency.

- Review of pre-planning and performance following an incident.

It also recommends the introduction of formal training sessions, seminars and site visits, conducted by employees with appropriate expertise, in order to build an awareness of the respective roles, operating procedures and problems faced by those involved.

As a general principle there should be operational cooperation between fire services and water undertakers to provide and secure water for firefighting. Water undertakers' expertise is central to the process of assessing and predicting the extent to which the distribution system can provide water for firefighting purposes. Such provision is therefore a joint process balancing what might be required with what may be available and then agreeing any actions necessary to fill the gap – if one exists. Particular attention should be paid to those potential incidents that carry the greatest risk and might demand substantial water resources for firefighting. Careful pre-planning, both to determine the likely water requirement and to identify available sources, is essential. These issues are considered in some detail in Chapter 7.

A particular concern of the National Liaison Group, which is addressed in the guidance document, is that of the possible effect on water quality of fire service operations. Any firefighting or testing of apparatus has the potential to affect the chemical or microbiological quality of the water. The causes include disturbing sediment in the main by changes in the rate of flow or flow reversal, and negative pressures in the main which could suck in contaminated water from the surrounding soil. It is essential therefore that:

(i) The liaison process should make the Local Fire Authority fully aware of the implications of discoloured water incidents. A communication process should be agreed which provides the water undertaker with the opportunity to consider whether anything can be done operationally to minimise the risk of supplying discoloured water to its customers.

(ii) In considering requests for additional mains capacity the optimum combination for the provision of water for firefighting should be sought which minimises both the need for oversizing and the potentially adverse impact on microbiological quality caused by stagnation. Full consideration will need to be given to issues which relate to the testing of hydrants, training and the design and maintenance requirements of any fire service equipment which may have an impact on water quality.

(iii) Fire services and water undertakers should consider the value to be gained from the flow testing of hydrants and of other methodologies for establishing potential flowrates which do not involve the risk of water discolouration and, as such, should preferably be used.

(iv) Service reservoirs are not used directly for firefighting purposes. However, following risk assessment, it may be necessary to consider the installation of a fire hydrant at the boundary of the site.

(v) Cross contamination of mains water with contaminated water is avoided. It is, therefore, important that equipment is adequately maintained, e.g. non-return valves on pumps.

(vi) Any actions which create sub atmospheric pressures in the mains should be prevented.

The introduction of additional legislation, to establish and enforce the responsibilities of water undertakers with regard to the water requirements for firefighting purposes, is a possibility.

4.3.3 Industrial Risks

There may be isolated patches of high risk in a predominantly low-risk area. In these cases, fire authorities usually advise the owners to install some form of adequate water supply for firefighting, e.g. reservoirs, underground tanks.

In many industrial premises large quantities of water will be required for processes carried on in the plant. Fire authorities usually have discussions with the owners and agree on the amounts they can take for firefighting. However, under Section 58 of the Water Industry Act 1991, developers, at their own expense, may require water undertakers to install and maintain hydrants specifically for fire-fighting purposes. Whatever the situation in these cases, it should be made quite clear to the owners that the fire authority cannot be expected to meet the expense of providing water supplies for special premises out of all proportion to the remainder of the risk in the area. The fire authority should advise such premises to install equipment in accordance with BS 5306: 1983 "Fire extinguishing installations and equipment on premises, Part 1: Hydrant systems, hose reels and foam inlets".

4.4 Pressure and Flow in Mains

It should be noted that, because of the variation in the demands from customers, flow and pressure in the mains will vary according to the time of day, day of the week and the time of year (see Figure 4.2) so that the quantity of water available for fire-fighting may also vary. The standing pressure at a hydrant, i.e. the pressure in the main when water is being taken only for normal domestic purposes, is not in itself an accurate guide to how much water will be available in an emergency situation. When firefighting water is being drawn the flowrate, and hence the loss of pressure due to friction (see Chapter 1), may, particularly if the main is of small diameter, be considerably greater than normal with the result that the pressure in the main will fall. The reduction in pressure depends on:

(i) the diameter of the main.
(ii) the condition of the main internally (which affects its diameter and roughness).
(iii) the length of the main.
(iv) the amount of water being drawn from the main.

For a small diameter main the internal pressure may reduce to not much more than atmospheric, even when only one hydrant is opened, so that attempts to obtain more water by opening neighbouring hydrants would be futile. To illustrate the point let us calculate the loss of pressure due to friction (P_f) when 1000 l/min (L) are drawn from a 100mm main (d) which is 200m in length (*l*). An appropriate value for the friction factor of a corroded main is of the order of 0.01, so the loss of pressure is given by:

$$P_f = \frac{9000 f l L^2}{d^5}$$

$$\text{i.e. } P_f = \frac{9000 \times 0.01 \times 200 \times 1000 \times 1000}{100 \times 100 \times 100 \times 100 \times 100}$$

i.e. Loss of pressure
= 1.8 bar

Thus, even at the relatively modest flowrate of 1000 l/min, it is quite possible that most of the available pressure in the main is used to overcome friction in the main itself. On the other hand, provided there is no constriction in it, a large main of 200mm or more will probably be able to supply several pumps before the pressure falls substantially (a similar calculation shows that, for the same flowrate, the pressure loss in a 200mm main is less than 0.1 bar) and consequently the location of these larger mains, especially those upstream of pressure reducing valves, and the hydrants on them is an important aspect of pre-planning.

Once the standing pressure is used to overcome friction in the main no further increase in flow will be possible no matter how many hydrants are opened.

Tuberculation and corrosion in mains will reduce their effective diameter and cause internal roughness which, together with the bends and fittings such as meters, will increase frictional loss. This is one of the main reasons why hydrants fitted on two different mains of the same diameter in the same pressure zone can give very different rates of flow. A knowledge of the capacity of every main is an important aspect of planning efficient fire protection of an area and consequently it may be necessary to conduct flow tests from hydrants, particularly those on long mains of small diameter. One proposal, under consideration at the time of writing, is for hydrant plates to give information about the flowrates available from the main rather than the main diameter.

4.5 Special Fire Mains

In some countries the system of special fire mains is employed. Such a system has the advantage that it can be planned solely to provide an adequate water supply for all fire risks at a suitable pressure and that untreated water may be used. Such mains are rare in this country, and only exist in special cases such as a large factory in an area where the public supply is not adequate to cover the special risk, and in certain docks. Sometimes these mains are fed from a river or canal or some other untreated water source, and in such cases it is essential to ensure that no cross connection is made during firefighting operations between foul water mains and potable water supplies. In other cases, special fire mains use water from the undertakers' supply.

All special fire mains must be fitted with British Standard hydrants.

4.6 Hydrants

4.6.1 Statutory Requirements

Section 57 of the Water Industry Act 1991 requires the water undertakers to allow water to be taken from their mains, by any person, for firefighting purposes and (at the fire authority's expense) to provide hydrants where the fire authority require them and maintain such hydrants in good working order.

Section 58 of the Act requires water undertakers, at the request of the owner or occupier of a factory or place of business and at their expense, to fix fire hydrants for the purpose of firefighting only.

For Scotland similar provisions are made in the *Water(Scotland) Act 1980.*

Although, with the exception of specially requested hydrants, the cost of installing, maintaining and renewing them is borne by the fire authority, the water undertaker is entitled to allow other individuals or concerns to use fire hydrants, and occasionally damage is caused in this way. Under Section 14(3)(b) of the Fire Services Act 1947, when damage is caused to a hydrant as the result of any use made of it with the consent of the water undertaker when not used for firefighting or other purposes of a fire brigade, the fire authority shall not be liable for the cost of repairing or replacing the hydrant incurred as a result of the damage.

Section 14(5) of the same Act makes it an offence to use a fire hydrant without proper authority, or to damage or obstruct a hydrant otherwise than in consequence of using the hydrant with the consent of the water undertaker.

4.6.2 Siting and Fixing of Hydrants

When water mains are installed or changed, the plans submitted to the fire authority will show the intended route and size of the mains. The fire authority should identify likely fire risks, estimate the water requirements and then specify the number and position of hydrants accordingly. The water undertakers should be able to state the approximate flow of water from each proposed hydrant under most conditions. The fire authority will then be able to decide whether the flow from the hydrants near the fire risk is sufficient to cover the risk adequately and, if not, to plan for additional supplies. Existing hydrants are normally placed at intervals of between 90 and 180 metres, but there are a number of current recommendations, regarding their spacing, for different areas of risk. Discussions are currently taking place concerning the introduction of guidelines for hydrant spacing on new developments and on whether the developer should be made responsible for meeting the cost of installation, such a change would require new legislation. Any new spacing guidelines are likely to be decided on a risk assessment basis.

The water main is provided with a branch or T-piece to which the hydrant is attached either directly or with a short length of pipe inserted. The hydrant is situated in a chamber or pit of brickwork or other suitable material which is covered with a removable or hinged lid, usually of cast iron. The hydrant valve is generally contained in the same pit but it is still possible to find old hydrant installations where the valve is in a separate pit. Such installations, however, are being gradually replaced.

The introduction of pillar hydrants with their advantages of greater conspicuity and higher delivery rates is a distinct possibility for the future. Fitting them with a Storz coupling would facilitate direct connection to large diameter hose.

Many old installations have the hydrant positioned on top of the T-piece on the main in the roadway but most modern installations are either on mains running underneath the pavement or, where the main is in the roadway, on branch pipes which bring them under the pavement. This has the advantage of causing the minimum obstruction of the roadway when hydrants are in use, and obviates the hydrant pit and cover being subjected to the strain from the passage of heavy vehicles which may make the covers difficult or impossible to open. However, to lay or re-lay mains in existing highways means that mains and hydrants may be located in roadways due to the lack of space on pavements. Putting hydrants on branches to locate them in the pavement may result in a "water quality hazard" by creating dead legs and also increases costs.

4.6.3 Standardisation of Hydrants

Under the Fire Services Act 1947, the Secretary of State was empowered to make regulations for the standardisation of fire hydrants. At the time of publication all hydrants are being installed to British Standard 750:1984 but a new European Standard, which will include pillar hydrants, is currently being written. The British Standard does concern itself with certain details of design but is being amended to be more performance-based in some areas. Manufacturers are free to develop their own patterns, but it specifies the more important features of sluice valve and screw-down hydrants and also gives dimensions for the openings of the surface boxes for use with such hydrants.

The spindles of all British Standard hydrants are screwed so as to close when turned in a clockwise direction, and the direction of opening is permanently marked on the hydrant spindle gland. The cast iron parts of hydrants are treated with an approved rust-proofing process.

Hydrostatic tests which fire hydrants should withstand are also specified in the Standard, the manufacturer being responsible for carrying out such tests; the Standard also requires that hydrants shall be of a pattern which, when fitted with a standard BS round thread outlet shall be capable of delivering not less than 2000 l/min at a constant pressure of 1.7 bar at the hydrant inlet. **Note: this clause is concerned purely with the hydraulic efficiency of the hydrant and does not place an obligation on the water undertaker to ensure this pressure and flow are available from the main.**

The British Standard recommends minimum opening dimensions for hydrant boxes and specifies that hydrant covers should be clearly marked by having the words Fire Hydrant or the initials FH cast into the cover.

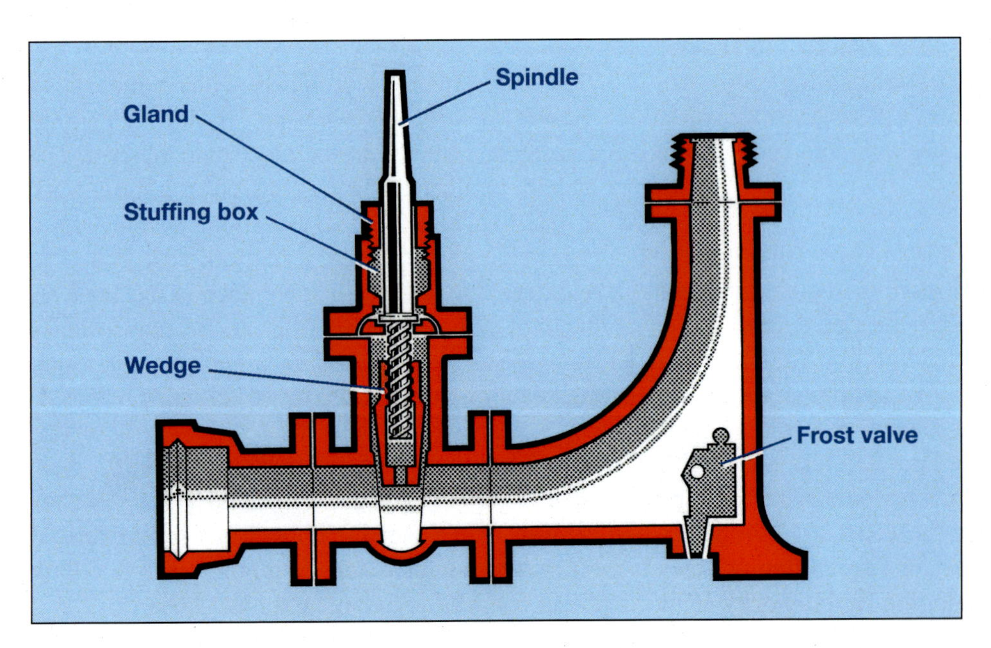

Figure 4.3 Section through typical sluice valve hydrant.

4.7 Types of Hydrant

The principal types of hydrant at present in use in this country are as follows:

4.7.1 Sluice Valve Hydrant

This type of hydrant (Figure 4.3) is not placed above the main, but alongside it on a short branch, the water flowing horizontally past the valve, and not vertically as in the screw-down type. It consists of three main castings, the inlet piece which is connected to the pipe, the sluice valve itself, and the duckfoot bend leading to the outlet. The opening and closing of the waterway is effected by means of a gate or wedge which may have gunmetal faces or be coated in a vulcanised rubber compound to effect a seal. Rotation of the spindle raises the wedge until it is clear of the waterway. The spindle passes through the valve cover by means of the usual gland and stuffing box.

This hydrant is hydraulically very efficient, and, when the valve is open, gives a full waterway with a negligible loss of pressure.

4.7.2 Screw-down Hydrant

This is probably the commonest type, being found in one or other of its forms in most parts of the country. It is attached directly to the main, which is provided at the chosen point with a vertical branch having a flange to which that of the hydrant is bolted. A mushroom type valve (obturator) (Figure 4.4) closes on a seating in the base of the hydrant body just above the inlet flange. The valve has a facing of

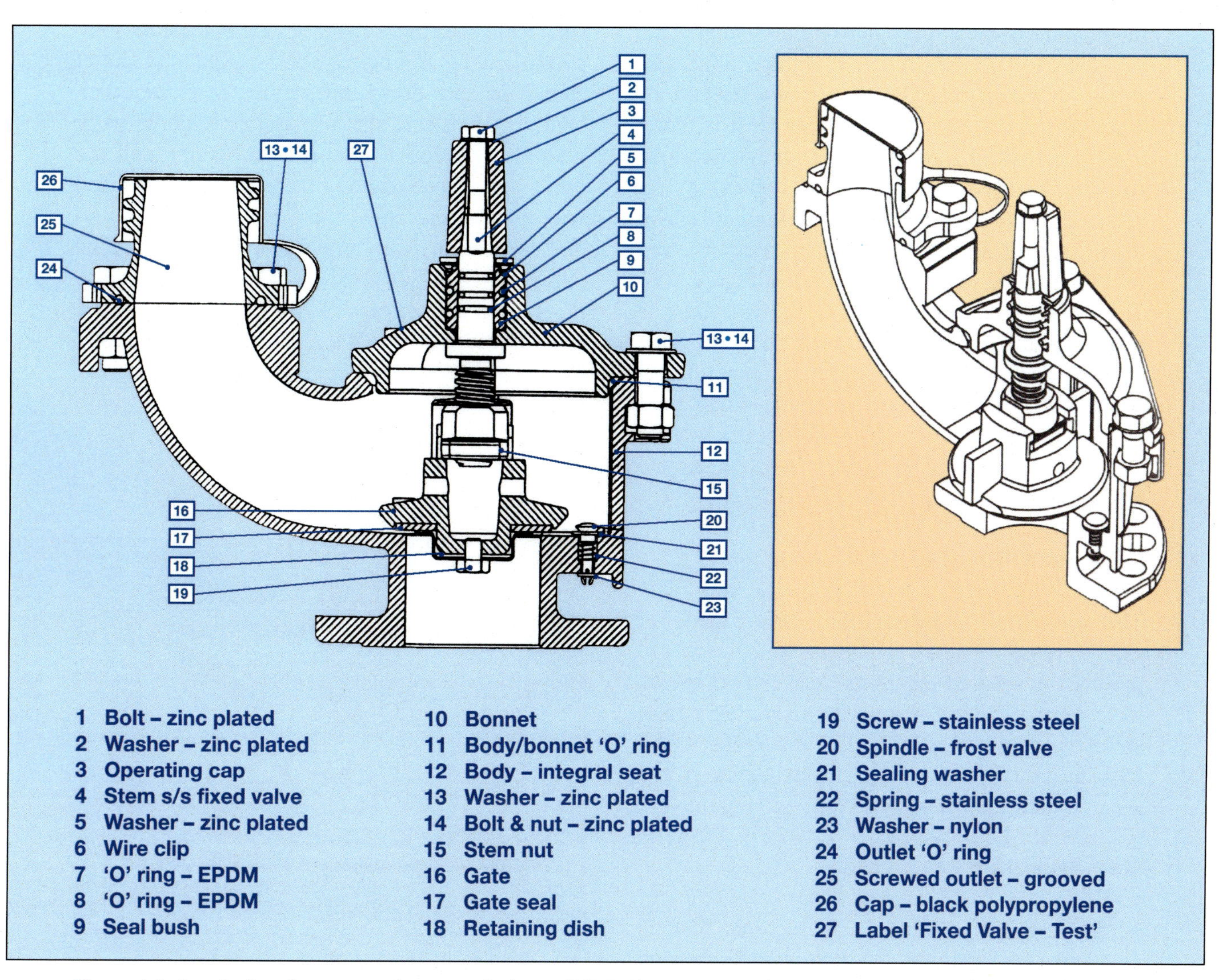

Figure 4.4 A typical modern screw-down type hydrant. (BS. 750) *(Diagram courtesy of Saint Gobain Pipelines plc)*

rubber or other suitable resilient material, while the seating may be of gunmetal or other suitable material. The valve is attached to the lower end of a screwed stem, and is lifted from its seating by the rotation of a spindle into which the stem screws until it is clear of the waterway. Guides are provided to ensure accurate seating of the valve. The outlet is normally bolted to the upper end of a bend leading from the valve seating. The hydraulic efficiency of this type varies greatly with the design of valve and outlet bends.

4.8 Hydrant Gear and Characteristics

4.8.1 Frost Valves

When the valve of a hydrant is closed after use, a certain amount of water is trapped in the body of the hydrant between the valve and the outlet. This is a source of danger in a cold climate, as it may freeze and so prevent the valve being opened, or may crack the hydrant body. Where frost is likely to be experienced, therefore, the hydrant should be fitted with means for draining off the water. The simplest and most common way of doing this is by a hole drilled in a small gunmetal plug screwed into the body at its lowest point (as in Figure 4.5). The disadvantage of this method is that water is constantly discharging through the hole while the hydrant is in use, though the loss is comparatively small. Older hydrants are frequently fitted with either a manually operated drain cock or an automatic drain valve which opens as soon as the hydrant is shut down.

4.8.2 False Spindles

The spindle of a hydrant is usually made of stainless steel, bronze or gunmetal, and, in order to protect the squared top from wear caused by the loose-fitting hydrant key, a cap known as a false spindle (or operating cap), made of a harder metal, such as cast iron or steel, is fitted over it and secured with a pin or screwed stud (see Figure 4.4).

4.8.3 Direction of Opening

Although British Standard 750 requires that all new hydrants should be made to open by turning the spindle anti-clockwise, in the past some were made to open clockwise. In any cases of doubt, both directions should be tried. All new British Standard hydrants and the underside of their pit covers are permanently marked with the direction of opening (Figure 4.5).

4.8.4 Outlets

The link between ground hydrants and the hose is provided by the standpipe, the base of which must connect to the outlet of the hydrant while the upper end provides the connection for the hose. All hydrants are now fitted with the standard round-thread outlet (Figure 4.3) as detailed in British Standard 750. However, older hydrants may have to be fitted with a round-thread adaptor.

Many hydrants on the larger sizes of main are made with a double outlet, to obtain a greater flow. On some older installations both outlets are controlled by a single valve so that one outlet cannot be shut down independently of the other. With this type it is also impossible to get to work with a single outlet without first blanking off the other, and a second line of hose cannot be added without shutting down the first line. These difficulties can be overcome by using a standpipe fitted with a rack valve. If, at the start of operations, such a standpipe is connected to the outlet not immediately needed, a second line can easily be brought into use when required by opening the rack valve.

Figure 4.5 The underside of a typical hydrant pit cover showing direction of opening the hydrant.

However, all recent double hydrants (Figure 4.6) have a separate valve for each outlet, so the flow to each can be controlled independently.

On old large capacity mains an outlet is sometimes found of the same diameter as the suction hose (Figure 4.7), enabling this to be connected directly to the hydrant. This outlet can be found in conjunction with a 65mm round-thread outlet as a double hydrant. Adaptors are also available which enable suction hose to be connected to a standard hydrant outlet. A suction outlet gives a significant increase in water output, especially when the normal running pressure is low, but if, as a consequence of large demand, the pressure in the main is reduced to below atmospheric, there is a strong possibility that, in the vicinity of leaks, escaped water may be drawn back into the main.

There are instances on record where polluted and toxic water has entered a main because of the negative pressure created by a large firefighting demand. Re-lined mains also present a problem because, under negative pressure, the lining may collapse.

4.8.5 Small Gear

Apart from standpipes, certain small gear for operating hydrants is carried by all firefighting appliances. This gear includes devices for lifting the hydrant cover and for operating the valve. The method of employment of such equipment is usually obvious.

4.8.6 Hydrant Pit and Cover

The pit in which the hydrant is contained has above it a metal frame and cover (Figure 4.7) which is flush with the roadway or pavement. British Standard 750 gives dimensions for all new hydrant pit covers and for surface box frames. British Standard 5306 Part 1:1983 gives details of typical forms of hydrant pit construction for either precast concrete sections or cemented brick-work, and recommends that the depth of the pit should be such that the top of the hydrant outlet is not more than 300mm below the surface of the road or pavement.

4.9 Hydrant Marking

Section 14(3)(a) of the Fire Services Act 1947 makes water undertakers responsible, at the expense of the fire authority, for causing the situation of every fire hydrant provided by the undertakers to be plainly indicated by a notice or distinguishing mark, which should be fixed to a conspicuously sited post erected for the purpose. (In cases where it is not possible to site an indicator post the indicator plate should be fixed to a nearby wall.) Lamppost mounting, where permission has been granted, is being increasingly used.

All recent hydrant plates conform to British Standard 3251:1976 and are made of vitreous enamelled mild steel, cast iron, aluminium alloy or plastics. The plates are yellow, with all characters and digits in black (Figure 4.8). Where required the yellow background may be of reflective material.

Figure 4.6 Double hydrant with each outlet controlled by a separate valve.

Figure 4.7 Hydrant fitted with suction outlet. Note the special draincock operated by a separate handle.

The Standard makes provision for four types of indicator plate, as follows:

> Class A plates: Hydrant indicator plates for general use except on roads of motorway standard.
>
> Class B plates: Hydrant indicator plates for use on roads of motorway standard.
>
> Class C plates: Indicator plates for emergency water supplies ('EWS').
>
> Class D plates: Indicator plates for meter by-pass valves.

Both Class A and Class B plates can refer either to single hydrants (indicated by the letter H) or to double hydrants (indicated by HD), and in each case the upper figure should denote the diameter of the main in millimetres while the lower figure gives the distance in metres between the indicator plate and the hydrant.

It is possible that on some older plates the two figures refer to inches and feet respectively but, as a general rule, it can be assumed that if the figure in the upper portion of the 'H' consists of two digits or more, the size of the main is in millimetres and the distance of the plate from the hydrant will be in metres. If, however, the figure in the upper portion has a single digit, the main diameter is in inches and the distance will be in feet.

The appropriate National Grid Reference may be marked at the top of a plate (above the figure indicating the main diameter).

The numerals may be removable, in which case they fit from the rear of the plate into slots, so that the figures are displayed through apertures in both the upper and lower portions of the letter H. Alternatively, hydrant plates may have raised or flush fixed letters and numerals.

Class B plates are larger than Class A plates, having additional space below the H or HD for the inclusion of suitable legend to indicate the location of the hydrant, the legend being at the discretion of the purchaser. This is necessary because fire hydrants are not normally located within the confines of motorways.

Class C plates are the same size as Class B plates and again the legend is at the discretion of the purchaser to suit the particular location concerned. Class C plates are intended for use in any appropriate situations, including motorways.

Where water is fed into industrial premises for business purposes through a meter it is a common practice for a by-pass to be fitted. In the event of the water supply on the factory side of the meter being required for firefighting, the meter can be by-passed, thus eliminating the frictional resistance through the meter. In addition, the water used for firefighting does not register on the meter. The position of the valve controlling the by-pass is generally marked by a standard Class D indicator plate (Figure 4.8) so that a firefighter can open the valve without loss of time. The figure on the plate indicates the distance in metres between the plate and the valve.

Due to cost factors, increasing use is being made of class A size plates to indicate open water supplies and by-pass valves with information displayed in the "main size" area of the plate.

4.10 Inspection and Testing of Hydrants

4.10.1 Background

In the past, hydrant inspection and testing was carried out based on the procedure described in Technical Bulletin 1/1994.

Tests were carried out on a regular basis though the frequency was left to individual Fire Authorities to determine.

When it had been thought necessary to conduct a flow test on a hydrant, because of uncertainty about the capability of a main to deliver sufficient water in an emergency situation, one of the specifically designed flowmeters described in Chapter 2 had been employed.

However, the Government's initiative on Best Value has recently prompted the CACFOA Benchmarking and Best Practice Project Team to challenge the need to conduct frequent testing of fire hydrants and, in particular, the stated need to

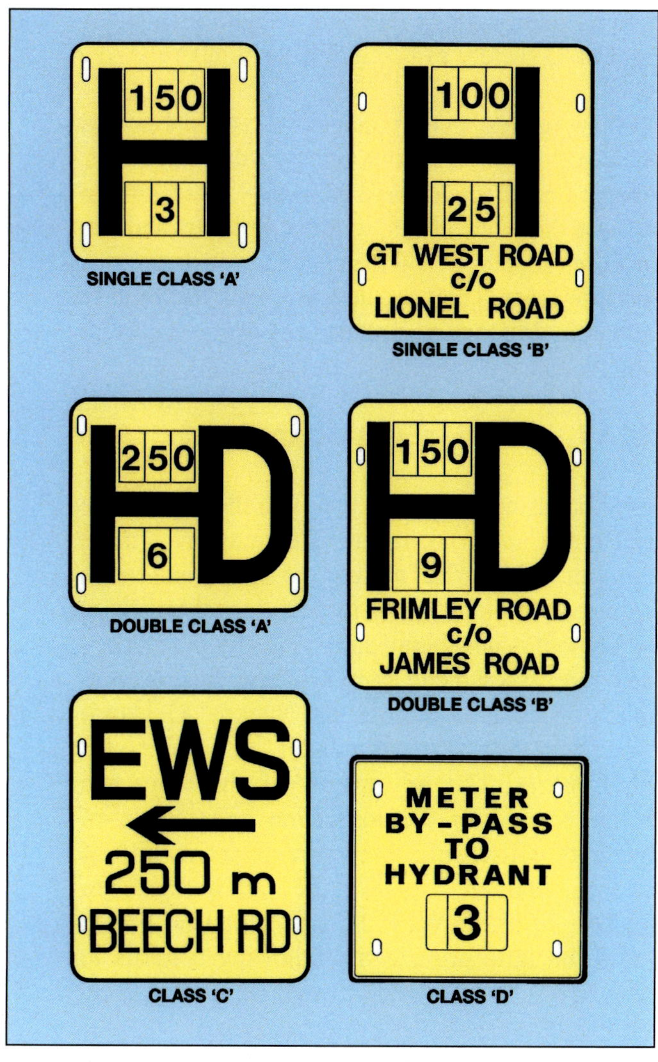

Figure 4.8 British Standard hydrant plates of Class A and Class B standard for marking single and double hydrants. Class C plates indicate the position of emergency water supplies. Class D plates denote the position of a meter by-pass valve.

fully open and close the hydrant valve. The arguments for and against such testing together with recommendations for a future testing methodology are summarised by the National Liaison Group in its publication "Guidance on Inspection, Testing and Abandonment of Fire Hydrants". The contents of this document are reproduced, largely in the original form, in paragraphs 4.10.2 to 4.10.7.

4.10.2 Reasoning behind the Fire Brigade Testing of Hydrants

The three popularly stated principal reasons for the fire service testing hydrants are:

- To ensure the hydrant is in fully operative condition.
- The hydrant is marked conspicuously to aid quick location.
- To increase the topographical knowledge of operational personnel.

The counter arguments are multifarious and include the following:

The maintenance of hydrants is not a legal responsibility of fire authorities, indeed Section 57 of the Water Industry Act 1991 (duty to provide a supply of water for firefighting) states under paragraph 3: 'It shall be the duty of every water undertaker to keep every fire hydrant fixed in any of its water mains or other pipes in good working order and, for that purpose, to replace any such hydrant when necessary". It is likely that, should fire brigades cease the inspection of hydrants, water undertakers would be obliged to take this task on and would pass on the costs of this work to fire authorities.

A hydrant flow test using one of the specially designed flowmeters described in Chapter 2, can only provide a 'snapshot' of the hydrant's capacity at the time of inspection – it provides no guarantee that a similar flowrate will be available the next time that it is used. The constantly changing pattern of demands for water by customers (see Figure 4.2) results in varying pressures in the system, negating the value of data taken at a specific point in time. Water undertakers, seeking to comply with ever more stringent water quality standards, are becoming increasingly concerned over the discoloration of mains water by the fire service testing hydrants. To a lesser degree, they are also concerned about the 'wastage' of water such tests involve.

The range of testing conducted by fire brigades varies enormously and may have little regard for guidance issued by manufacturers, but the often used method of 'cracking the valve' (to see that the main has water in it) is viewed by water engineers as having no engineering value. Indeed it is felt that this partial opening may cause more premature wear to the spindle and seat than if it was left in the closed position. In practice, fire brigades report more defects during testing than at other times and it is possible that the test process itself has caused many of the defects. The five known

U.K. hydrant manufacturers/suppliers have been consulted and none requires an annual wet test.

The requirement to provide adequate training for service hydrant test/maintenance personnel under the Highways Works Regulations removes valuable time and effort from fire service core business activities. The testing of hydrants may result in a higher incidence of accidents to fire service personnel and vehicles due to increased road mileage and the activity on the roadside. Increased fuel usage is both expensive and environmentally unfriendly as is the increased wear and tear on vehicles. Time spent by fire service personnel on hydrant inspection and testing is lost to other causes such as for example community fire safety education, training, etc.

4.10.3 Objectives and Issues for Consideration in Formulating Policy

The prime concern of water undertakers appears to be the impact of hydrant inspection, testing and flow testing activities by fire brigades on the quality of water in their distribution systems. The prime concern of fire brigades appears to be the validity of hydrant testing and inspection and the cost of repairs. Local discussion on the relative benefits of these activities may help undertakers and fire brigades to review their polices.

The various objectives of fire brigades and water undertakers in relation to fire hydrants are:

- To ensure that there is access to water for firefighting purposes.
- To maintain hydrants efficiently at minimum cost.
- To minimise, if not eliminate, the risk of disruption and discoloration of water supplies.

There is a balance of risk, cost and benefit to society in the continuing provision and maintenance of a hydrant on a distribution system. The fire brigades trade off the benefit of being able to have access to water in the event of a fire against the ongoing costs of providing and maintaining the hydrant. The water undertaker carries the risk of interruption or discoloration of supplies, but weighs them against its legal obligations and the benefit to society of being able to fight fires.

Both industries have an obligation to the public and their customers to ensure that they discharge their obligations effectively and efficiently. In seeking to achieve these objectives, water undertakers and fire brigades should review the content of their policies in these areas. They should test jointly the validity of continuing historic practice against their current obligations and common objectives.

4.10.4 Risk Assessment Approach

Risk assessment is a term that both the fire service and water undertakers are becoming ever more familiar with. The culture of the past which required the monotonous inspection and testing of equipment is now being superseded by the modern day risk assessment approach and this is very much encouraged through the National Guidance Document on the Provision of Water for Firefighting. The advantages of this approach being applied to the inspection and testing of hydrants will ensure that hydrants are monitored and maintained to meet the requirements of Section 57 of the Water Industry Act 1991 and at the same time provide the following benefits to both fire authorities and/or water companies.

- Significant reduction in hydrant repairs and maintenance budgets.
- Reduced administrative costs.
- Reduced risk of causing discoloration of drinking water.
- Resources redirected to more proactive tasks (eg community fire safety).
- Improved liaison arrangements between the organisations.

4.10.5 Recommended Future Hydrant Inspection and Testing Methodology

It is recommended that future inspection and testing of hydrants should consist of one of the three examinations:

- above ground
- below ground
- wet and pressure test

In line with the National Guidance Document on the Provision of Water for Firefighting, fire brigades and water undertakers are encouraged to move away from the flow testing of hydrants and use other methodologies. This is not seen as a test for maintaining a hydrant and, for the reasons highlighted in paragraph 4.10.2, there is little purpose to this test. As a result, the flow test is not included as one of the examinations of hydrants.

Above Ground Examination

This will involve a visual inspection of the hydrant frame, cover, surface surrounding the hydrant and the hydrant indicator plate. The period between inspections should be risk assessed and take into account such likely factors of area location and risk, hydrant position, age, material, previous history, etc.

E.g. For a hydrant situated in the pavement of a residential urban area free from vandalism, the fire brigade may determine to inspect on a 1–2 year basis, whereas a hydrant set in a country lane that has regular farm traffic driving over it may need inspecting every 3–6 months to ensure it is clear of mud, etc.

Below Ground Examination

This will involve the visual inspection of the hydrant pit and the hydrant itself. Defects which would affect the ability to deliver water for firefighting purposes or create a hazard should be reported immediately. The period between inspection should be risk assessed and take into account area location and risk, hydrant position, age, material, previous history, etc.

E.g. For a hydrant situated in the pavement of a residential urban area free of vandalism, the fire brigade may determine that an inspection should be carried out every 2–4 years, whereas a hydrant that regularly silts up may require inspecting every 6 months.

Wet and Pressure Test

The hydrant test is conducted by fitting a standpipe to the outlet and then partially opening the valve to allow a small amount of water to flow (equivalent to a domestic tap). A blank cap is to be fitted in the standpipe head, or the valve in the head closed and the hydrant fully opened. Whilst under pressure, all joints are to be visually inspected for signs of leakage and only those leakages that would impair the hydrant for firefighting purposes, or cause a hazard, should be reported to the water company. The hydrant is to be turned off without excessive force and the standpipe removed.

This test should only be carried out where there is reason to doubt the hydrant's integrity or that it is at an interval recommended by the hydrant manufacturer.

4.10.6 Recommended Procedure following use by Fire Brigades of Hydrants at Operational Incidents

Following use at an operational incident, the opportunity should be taken to record the hydrant number/location and to note any defects which would otherwise have been found during a hydrant examination highlighted above. This will reduce the time fire brigades have to spend inspecting hydrants and will provide a record of when the hydrant was last used. This may be particularly important should a third party have used or damaged the hydrant and the fire brigade receive an invoice for the hydrant repair. Water undertakers would also welcome being notified where a hydrant has been used at an operational incident as it will aid their monitoring of usage and leakages.

4.10.7 Maintenance Costs

Cost is driven by a number of factors, but includes: the number of hydrants in a fire brigade area, inspection and testing policy, direct maintenance practices and administrative procedures. There will be a wide range of local circumstances that contribute to current practice across the country. The following points may help water undertakers and fire brigades to question and, therefore, improve current practice through liaison and agreement.

Number of Connected Hydrants

- Review through a risk assessed approach the number of hydrants required for a given area.
- Review policy for the provision of new hydrants.
- Consider a phased programme of abandonment spread over a number of years.
- Consider a policy to abandon hydrants as an alternative to repair.
- Consider opportunistic abandonment of hydrants during water undertaker mains renewal or rehabilitation schemes.

Maintenance Practices

- Consider standard repair packages.
- Allocation of repair tasks between water undertakers and fire brigades.
- Economies of scale of operations with other water undertakers or brigades.
- Economies of scale through standardisation or joint purchasing.
- Reduce inspection and testing frequency based on a risk analysis.
- Timing of inspection and repair throughout the year to match available resources.

Administrative Procedures

- Preparation of accounts, e.g. monthly, quarterly, etc.
- Formulation of fixed prices for standard repairs.

4.11 New Roads and Street Works Act 1991

Sections 65 and 124 of this Act require anyone carrying out work under the Act to do so in a safe manner as regards the signing, lighting and guarding of the works. This has implications for the fire service when hydrant testing and inspections and other minor works by arrangement with water authorities are undertaken. The regulations emphasise the importance of making sure that all workers engaged in street and road works are safe and that drivers and pedestrians are made aware of any obstructions well in advance. Details of how to comply with the regulations are explained fully in "Safety at Street Works and Road Works – A Code of Practice", available from TSO.

Chapter

5

Chapter 5 – Pumps and Primers

Introduction

Essentially, a pump is a machine, driven by an external power, for imparting energy to fluids. Power may be provided by the operator, as in hand pumps and hand operated primers, or by coupling the pump to a suitable engine or motor. This Chapter is principally concerned with the latter, although it deals with the general principles of all types.

The Fire Service has come to rely mainly on centrifugal pumps and these, together with the primers necessary to get them to work from open water, are described in detail. The first part of the Chapter looks at the principles of operation of all the types of pump which may be used by brigades and then progresses to a more detailed examination of centrifugal pumps, both vehicle mounted and portable, and their cooling and priming systems. Chapter 6 deals with practical pump operation.

5.1 Operating Principles of Non-Centrifugal Pumps

Non-centrifugal pumps used by the Fire Service are based on one of two operating principles. These are:

5.1.1 Positive Displacement Pumps

These usually have a reciprocating piston which makes an air- and liquid-tight seal with the cylinder in which it moves. The principle of operation of the pump is shown in Figure 5.1.

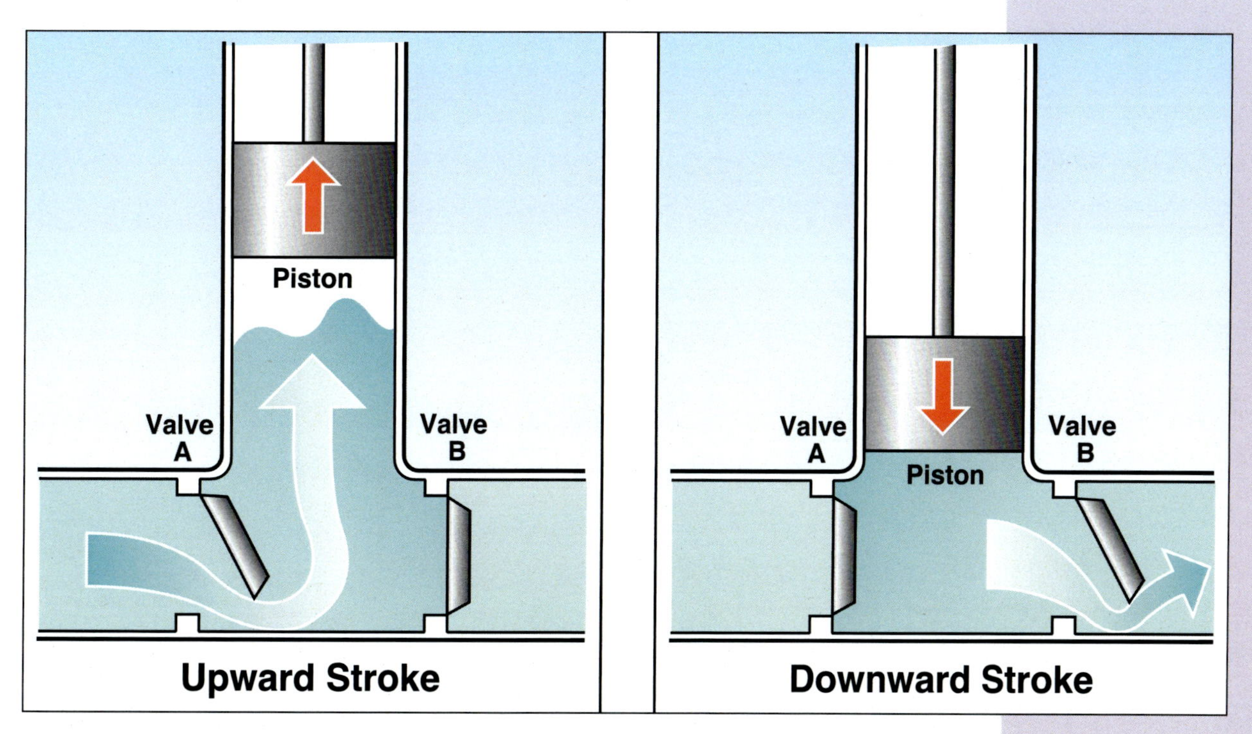

Figure 5.1 The principle of the reciprocating pump.

On the upward stroke of the piston a reduced pressure is created in the cylinder causing the inlet valve (A) to open and the outlet valve (B) to close, so that air or water is drawn into the cylinder through the inlet valve. On the downward stroke the inlet valve closes and the contents of the cylinder are forced, under positive pressure, through the outlet valve.

Application of the reciprocating pump principle, though with a more efficient valve arrangement, may be found in the stirrup pump (described fully in the current Book 3 of the Manual of Firemanship: Hand pumps, extinguishers and foam equipment) and in certain fire pump primers which will be described later in this Chapter.

5.1.2 Ejector pumps

There are several varieties of ejector pump in use in the Fire Service. This Section refers to those used for pumping water. The exhaust gas ejector primer, which works on a similar principle, is dealt with separately in Section 5.4.2.

The principle of operation of an ejector pump, which is referred to in Chapter 1, may be more fully appreciated by reference to Figure 5.2.

Water under pressure from another pump (the propellant) emerges in jet form from a small internal nozzle and enters the delivery tube via an opening known as the throat. The narrowest part of the throat is slightly larger than the orifice of the nozzle and is separated from it by a gap which is open to the surrounding fluid. As the jet passes the gap and rapidly expands, the consequent fall in pressure at the throat causes surrounding water at atmospheric pressure to join the stream.

Figure 5.3 shows, in diagramatic form, a typical example of an ejector pump which may be suspended above the water line and propelled by water supplied via a standard instantaneous coupling.

Ejector pumps are light and easy to handle, and can be used in situations where it is undesirable to use conventional pumps, e.g. due to hazardous fumes. Furthermore, their operation is unaffected by an oxygen-deficient atmosphere which would cause an internal combustion driven pump to stall. Once set up they require little or no attention except the removal of debris that may have collected in the suction strainer. The primary pump supplying the water to the ejector pump can be placed in a convenient and safe position e.g. away from smoke and other hazards.

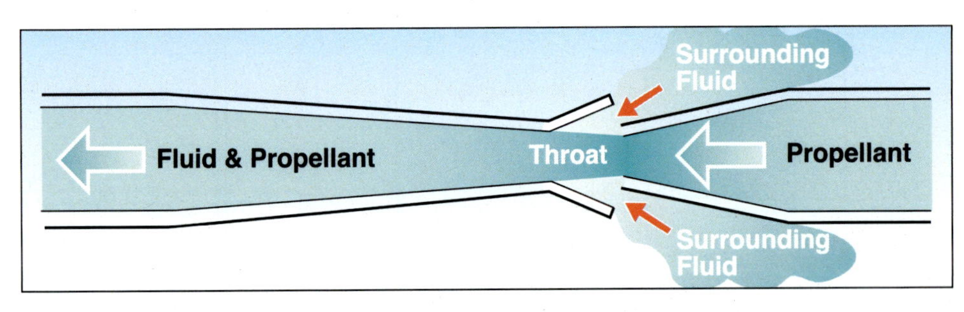

Figure 5.2 The operating principle of the ejector pump.

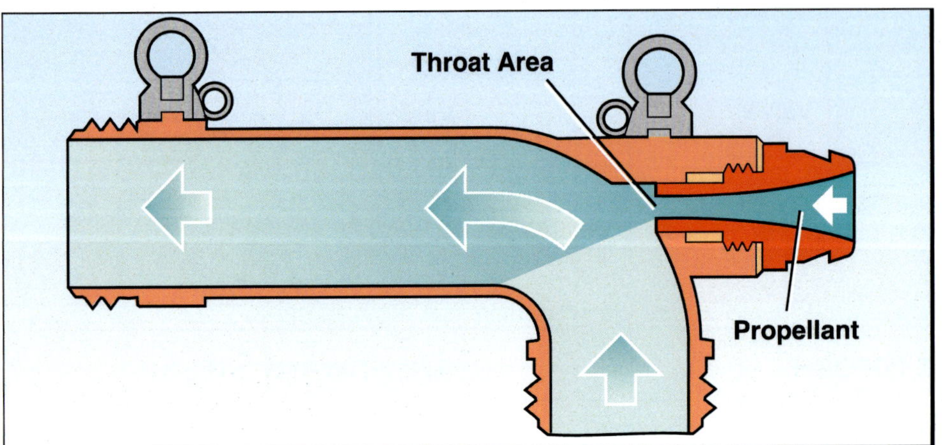

Figure 5.3 Diagram of a typical suspended type ejector pump.

The quantity of water lifted by an ejector pump will vary according to:

(i) the height of the ejector above the water level;

(ii) the height of the discharge point above or below the ejector.

If the discharge point is above the ejector, the output will be reduced appreciably, and it is therefore important to keep the discharge outlet as low as possible. The actual amount of water pumped out is normally the difference between the input and the total discharge. It is possible for the water being pumped out by the ejector pump to be recirculated via the primary pump, thereby providing the necessary propellant for the ejector pump. The surplus water may then be discharged by other primary pump deliveries.

Another type of ejector pump is the submersible type, examples of which are shown in Figures 5.4 and 5.5, which may be used for pumping out water from depths greater than the maximum suction lift. The pump may rest on its base, which is the suction inlet and is fitted with a low-level type of strainer. The body of the pump has an inlet for water from the primary pump, and a discharge outlet. The example shown in diagrammatic form in Figure 5.5 has a two-stage ejector nozzle incorporated and its performance characteristics are shown in Figure 5.6.

Example 1

What will be the total amount of water discharged from the submersible type ejector pump when operating at a lift of 8 metres if the input is 864 litres per minute at 7 bar?

Inspection of the appropriate graph indicates that the total output will be approximately 1875 litres per minute so that the amount of water pumped out will be about 1000 litres per minute.

The maximum possible lift for the two-stage type of pump is about 18m.

Figure 5.4 A typical small hosereel propelled submersible ejector pump. (Courtesy of West Midlands Fire Service)

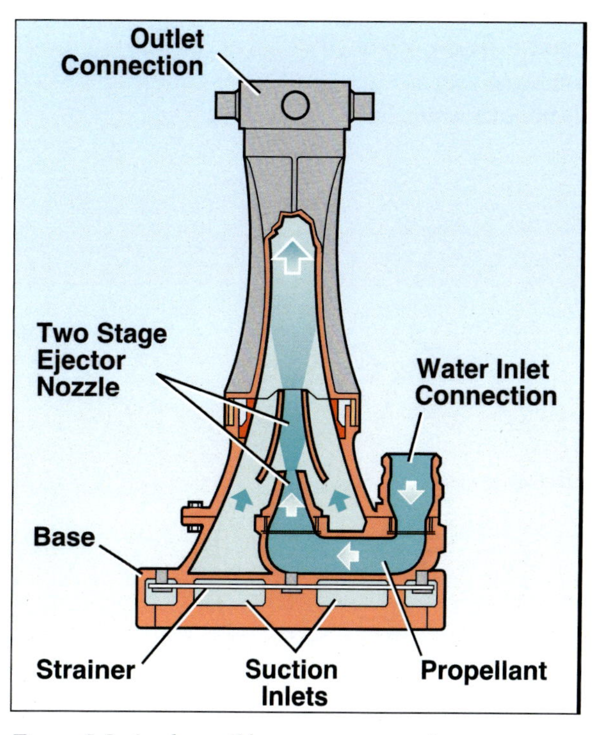

Figure 5.5 A submersible type two-stage ejector pump in diagrammatic form.

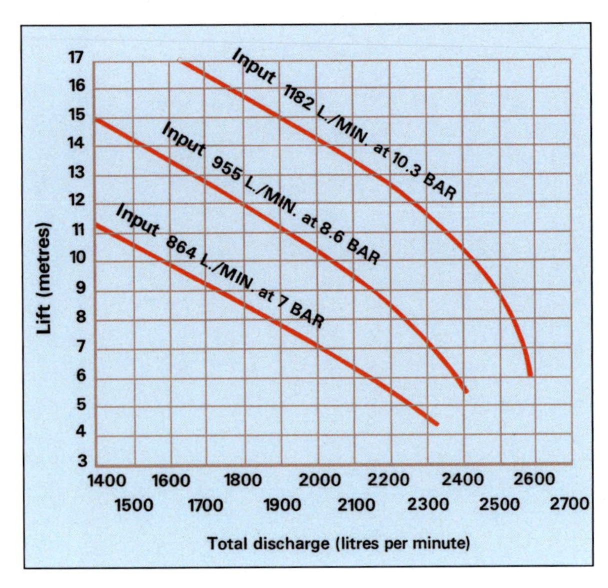

Figure 5.6 Performance characteristics.

5.2 Operating Principles of Centrifugal pumps

Centrifugal pumps are the most widely used for firefighting. They are unable to pump gases (and therefore have to be primed), have no valves, pistons or plungers and do not work by displacement. Instead they make use of *centrifugal force* (i.e. the force which a rotating body experiences tending to make it fly away from the axis of rotation) in much the same way as a spin dryer uses centrifugal force to remove water from wet clothes.

A centrifugal pump consists essentially of a spinning circular metal casting with radial vanes, called the impeller (Figure 5.7), enclosed in a casing. Water at the centre of the impeller is thrown outwards by centrifugal force as the impeller rotates and discharged at the periphery thereby causing a partial vacuum to be created at the centre. This causes more water to be forced into the impeller from the supply source so that flow from the centre of the impeller to its periphery is continuous.

The action of the impeller in thrusting water outwards naturally creates considerable turbulence and friction and, as these factors cause some of the power used to drive the pump to be wasted and so reduce pump efficiency, it is important to minimise their effect. This is achieved by careful design of the casing, and possibly by the introduction of a system of guide vanes called a diffuser, to ensure that flow is, as near as possible, streamlined. Figure 5.8 shows simplified diagrams of a centrifugal pump with and without guide vanes. As water moves away from the centre of the impeller, and travels on its way to the outlet, the area of cross-section of the path along which it passes increases, thereby causing the velocity and kinetic energy of the water to decrease but with a consequent increase in pressure. With many pumps a further increase in cross-sectional area of the channel occurs in a snail shaped part of the casing called the volute.

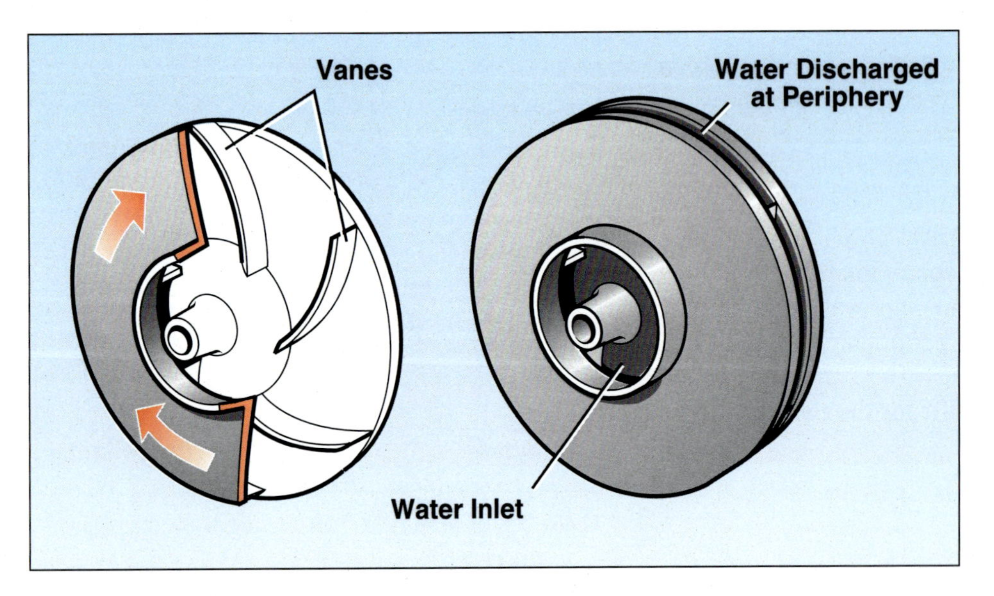

Figure 5.7 The construction of a typical impeller.

Figure 5.8 *Simplified diagrams showing the operation of a single stage centrifugal pump. Left: without guide vanes. Right: with guide vanes.*

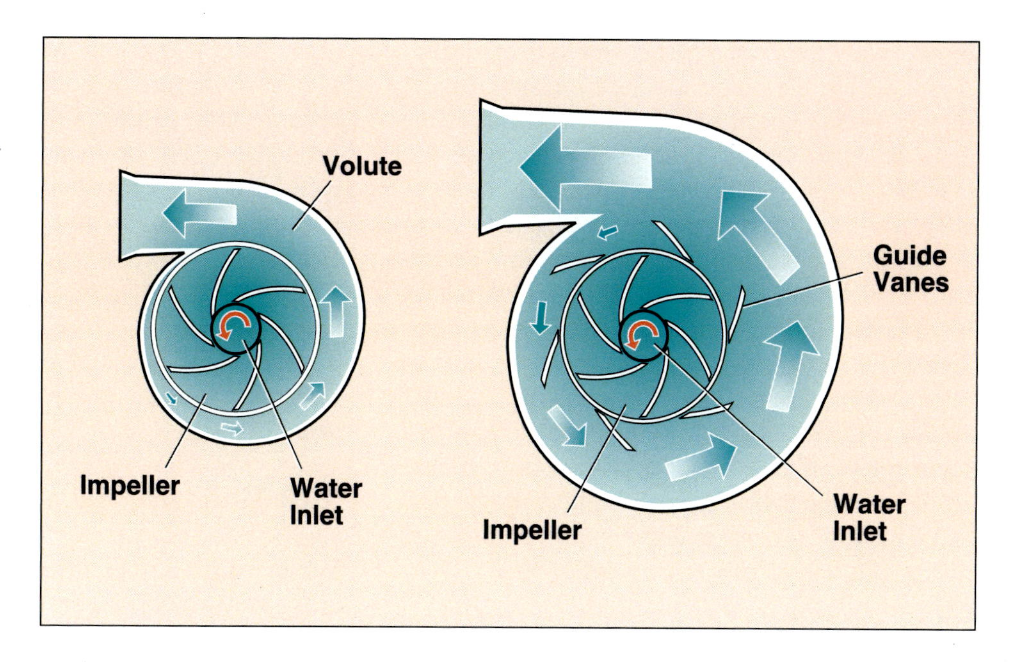

Figure 5.9 A typical set of pump characteristics for different lift conditions as presented in the manufacturer's brochure.

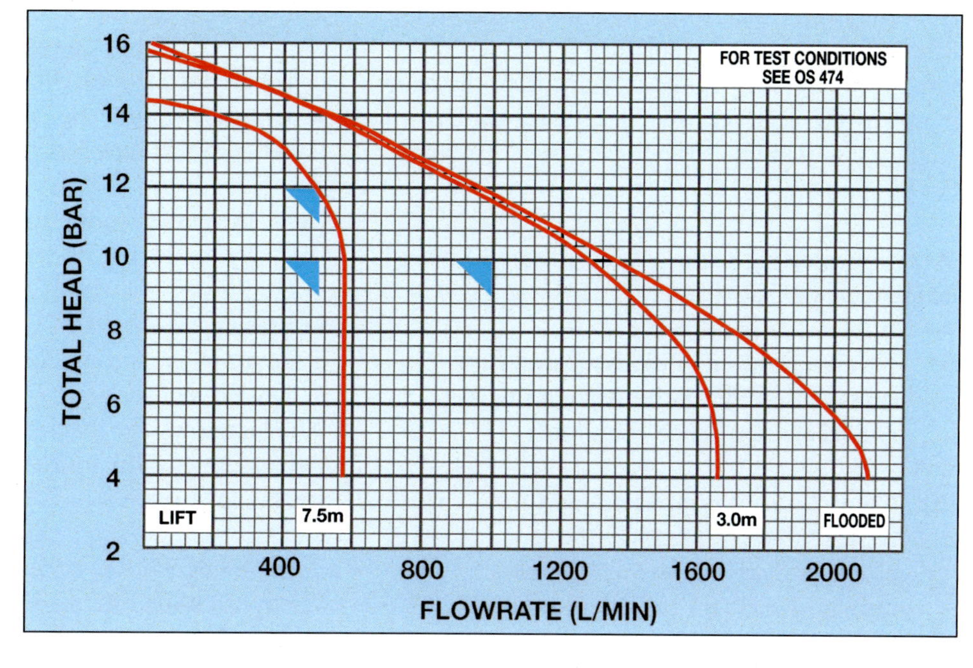

The changes in energy which the water undergoes as it passes through the pump and the subject of pump efficiency are both discussed in Chapter 1.

5.2.1 Pump Characteristics

For effective use of a fire pump in an emergency situation and particularly in pre-planning, it is important for firefighters to know the maximum output in litres per minute (l/min) of which the pump is capable at any given operating pressure. This information is best presented graphically by means of what is known as the pump characteristic (Figure 5.9) for any given pump/engine combination and is usually available, in this form, from the pump manufacturer.

Although it is possible to produce a characteristic for any given engine speed, the most useful, for practical purposes, is the one obtained at the maximum throttle setting (but probably with an upper limit imposed on engine r.p.m.) which therefore indicates the limits of performance for that particular pump.

Inspection of the characteristics shown in Figure 5.9 indicates that the pump develops maximum pressure when the discharge from it is zero, that the pressure decreases as the discharge increases and that the maximum discharge is obtained when operating against minimum pressure. In Chapter 3 it was explained that suction lift affects pump performance, so it is important, when data on performance is presented, to specify the relevant lift conditions. Figure 5.9 shows three characteristics – for lifts of 7.5m, 3.0m and zero (for example with the compound gauge reading zero as might be the case in a water relay). If only a single characteristic is presented, or if the data is given in a different way, such as by quoting the maximum output at a particular pressure, it will usually be for a lift of 3m.

It is because the pressure available from a pump (for any given flowrate) decreases with increasing lift, that, in a water relay supplied from open water, it is recommended that the distance from the base pump to the first intermediate pump should be slightly less than the spacing between the remainder of the pumps.

When interpreting pump characteristics it should be appreciated that:

(i) No combination of pressure and flow represented by a point *beyond* the characteristic line is achievable.

(ii) Combinations of pressure and flow represented by points *on* the characteristic line are achievable only at maximum throttle/r.p.m.

(iii) Combinations of pressure and flow represented by points *within* the characteristic line are achievable at reduced throttle settings.

It is common practice for manufacturers to quote pump performance in terms of the maximum output available at a specific pressure – typically 7 bar – so, for example, a 2250 l/min pump is described as such because it is able to achieve that flowrate when operating at this pressure. However, it is becoming more common, because of proposed European standards, for manufacturers to quote pump output at an operating pressure of 10 bar. The pump whose characteristics are shown in Figure 5.9 is marketed as a 10/10 because (at 3m lift) it is easily able to deliver 1000 (10 × 100) l/min at 10 bar. It might equally well have been described as a 1400 l/min pump because it is easily able to deliver this amount at 7 bar.

Care should be exercised, when comparing pumps on the basis of their outputs, to ensure that these outputs are quoted at the same operating pressure.

5.2.2 Multi-stage pumps

Pumps with a single impeller, as described above, are capable of developing pressures of anything up to about 20 bar, depending on the particular design and the flowrate required. If higher pressures are required, for the operation of high pressure hosereels for example, there are two methods by which they might be achieved with a single impeller of this type:

(i) by increasing the speed of the impeller;
(ii) by increasing its diameter.

Increasing the speed of the impeller can only be achieved by increasing the speed of the engine and it may be neither practicable nor desirable to do this. Increasing the diameter is comparatively inefficient and would make the pump more bulky. Hence, to achieve a high outlet pressure, it is better to use a multi-stage pump, i.e. a pump with two or more impellers in series. The impellers are driven by the same rotating shaft with water fed from the periphery of the first impeller to the entry to the second etc., so that, neglecting friction losses, the pressure increasing ability of the centrifugal pump is applied a number of times. If the impellers are of the type already described, then several stages are required to achieve the high pressures needed for hosereel operation and, although pumps of this type are in use, most fire service pumps consist of only two stages and use a different type of impeller for the second stage. This type of impeller, called the **regenerative** (or peripheral) type, is described in the next section.

Practical multi-stage pumps have the important advantage that water may be discharged at relatively low pressure after passing through the first stage only, or at high pressure (up to about 55 bar) after passing through subsequent stages. It is also possible to deliver water at both high and low pressure simultaneously.

Vehicle mounted pumps are normally two-stage and designed to be driven through the power take-off of the engine of a stationary appliance. The speed at which the pump can run is, therefore, conditioned by the capability of the engine from which it is taking its power. Pumps designed for high speeds, however, may have a speed-increasing gear built into them or at the power take-off.

Portable pumps are invariably single stage.

5.2.3 Regenerative (peripheral) pumps

A two-stage pump consisting of two identical impellers would develop a second stage working pressure of approximately twice that of the first stage, and, for that second stage pressure to be high enough for hosereel operation, very high engine speeds would be required. This, in turn, would mean that the first stage pressure could be dangerously high for use with normal low pressure firefighting hose and branches. Therefore in order to achieve a high second stage pressure at a moderate pump speed a different sort of impeller, called a regenerative type, is employed for that stage.

At least one major pump manufacturer has, for safety reasons, a design policy for firefighting pumps such that, at any given engine speed, the first stage pressure cannot exceed one quarter of the second stage pressure.

The use of the regenerative impeller not only results in a saving on the wear and tear of the moving parts but makes a speed increasing gear unnecessary. A regenerative impeller is shown in Figure 5.10 and its operating principle may be understood by reference to Figure 5.11.

The regenerative impeller has, at its periphery, a ring of guide vanes and is enclosed in a casing which fits closely around the central part of the impeller but leaves a channel around the part containing the vanes (see Figure 5.11). Water enters this channel via an inlet, drops to the base of the guide vanes, and is then thrown outwards between the vanes by centrifugal force. It then moves to the base of the vanes again and the process is repeated. This happens several times whilst the water is dragged round in a circle by the rotating impeller, until it finally reaches the outlet. The water therefore follows a spiral path around the channel. The cumulative effect is to develop a sufficiently high

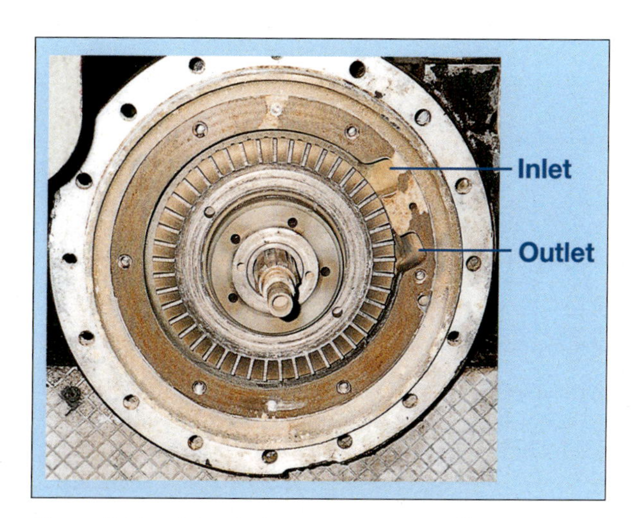

Figure 5.10 a (above) A regenerative impeller.
(Courtesy of Fire Service College)

Figure 5.10 b (right) Components of a two-stage pump.
(Courtesy of Hale Products Europe)

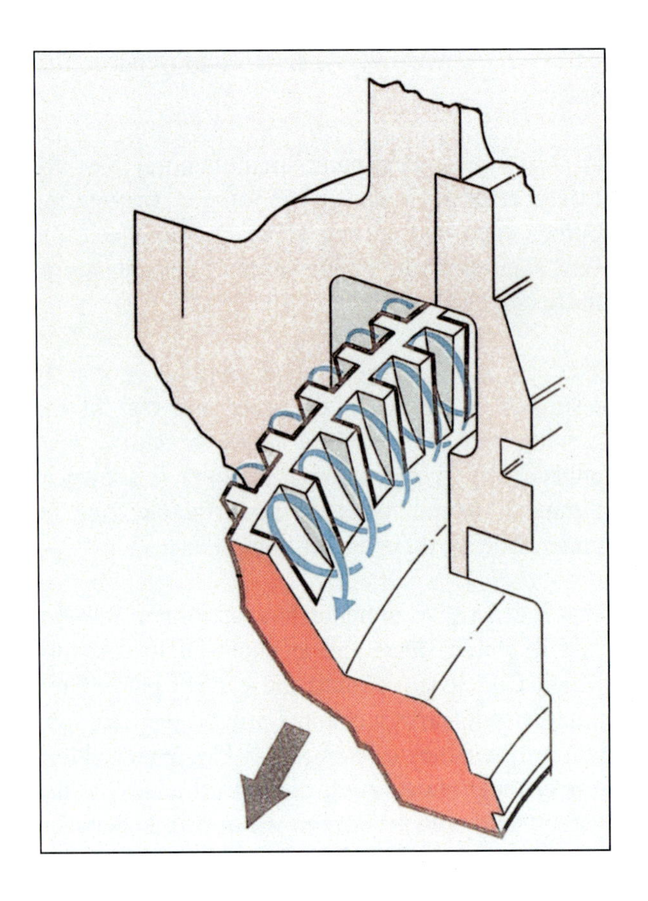

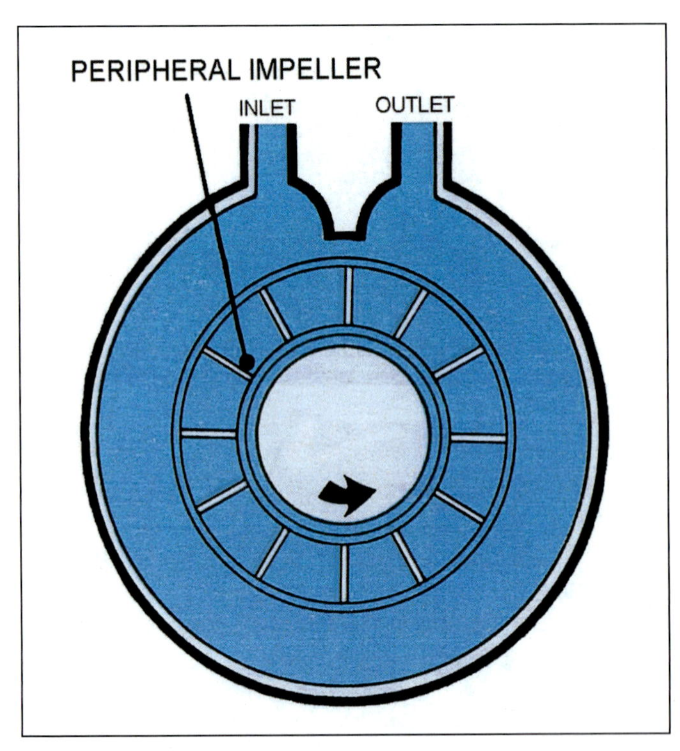

Figure 5.11 (left and above) Diagram showing the operating principle of the regenerative pump.

pressure for hosereel operation and water-fog production but at a pump speed of the order of 3000 rpm. The characteristics, at given engine speeds, for a typical two-stage pump are shown in Figure 5.12.

5.3 Vehicle Mounted Fire Pumps

5.3.1 Examples of vehicle mounted pumps

At the present time, the specification for vehicle mounted pumps is that of JCDD/3/1 as amended in October 1985. However, a draft European Standard, prEN 1028, is in the course of preparation and eventually a new JCDD specification will be based upon it.

JCDD/3/1 requires that the low pressure stage of a multi-stage pump should be capable of delivering a minimum of 2270 l/min at a pressure of 6.9 bar and, simultaneously, the high pressure stage a minimum

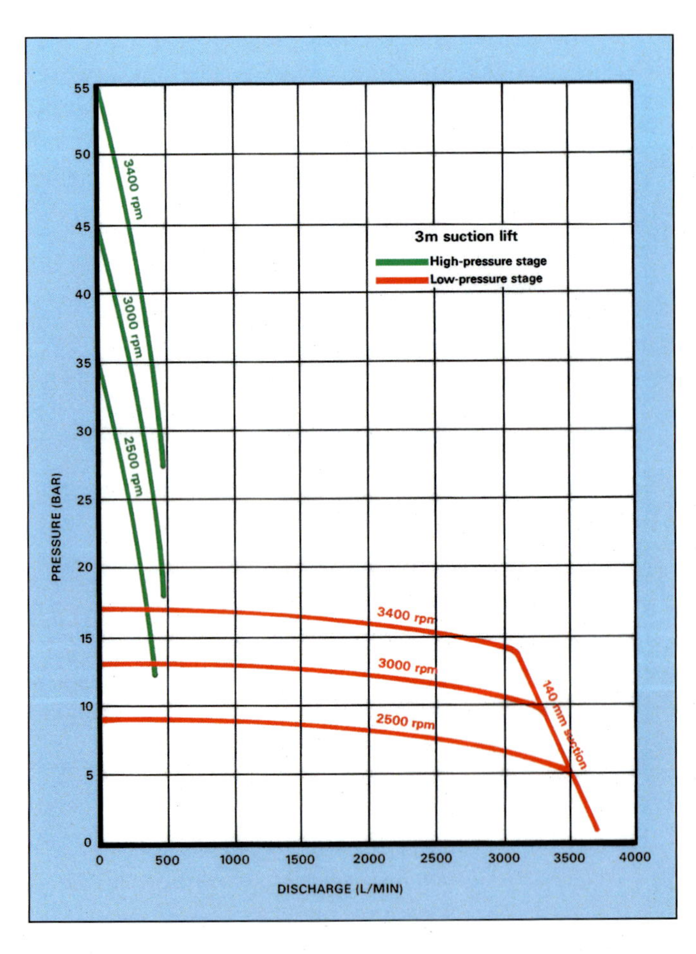

Figure 5.12 (right) Performance characteristics for a typical two-stage vehicle mounted pump.

of 227 l/min at 24.1 bar at an engine speed less than the maximum recommended by the manufacturer for continuous duty running.

Vehicle mounted pumps, because of the need to provide a choice of pumping pressure, are almost invariably of the multi-stage type. Figure 5.13 shows a sectioned diagram of the recently introduced GODIVA World Series two-stage type of pump. It has a normal centrifugal impeller as the first stage and a regenerative type for the second. Figures 5.14 (i) to 5.14 (viii) illustrate the various modes of operation of the pump.

Whenever HP operation is required the Change Over Valve must be closed, otherwise water from the HP stage will simply escape back to the LP stage.

Under conditions of no discharge it is important that a limited escape of water takes place otherwise overheating will occur.

Figure 5.14 (i) represents the LP (i.e. change over valve open) mode but with no discharge. Once a temperature in excess of about 42 degrees Celsius is attained the thermal relief valve (TRV) opens to allow a certain volume of water to escape and a corresponding amount of cold water to enter.

Figure 5.14 (ii) shows the change over valve closed and no discharge from either delivery. If the pump speed is high enough to cause pressure in excess of 55 bar in the HP stage the pressure relief valve (PRV) opens and allows circulation from the HP to the LP stage. Again the TRV will open, if necessary, to prevent excessive temperatures.

Figure 5.14 (iii) shows HP only operation. The TRV remains closed but the PRV will open if excessive pressure is attained in the HP stage.

Figure 5.14 (iv) shows simultaneous HP and LP operation. The PRV and TRV remain closed.

Figure 5.13 The Godiva World Series two-stage pump. *(Courtesy of Hale Products Europe)*

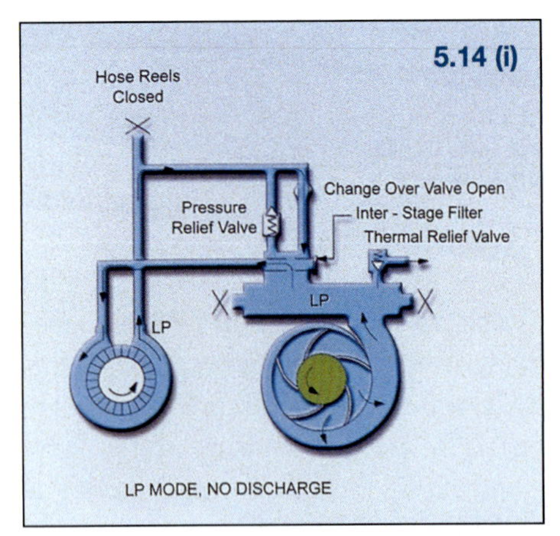

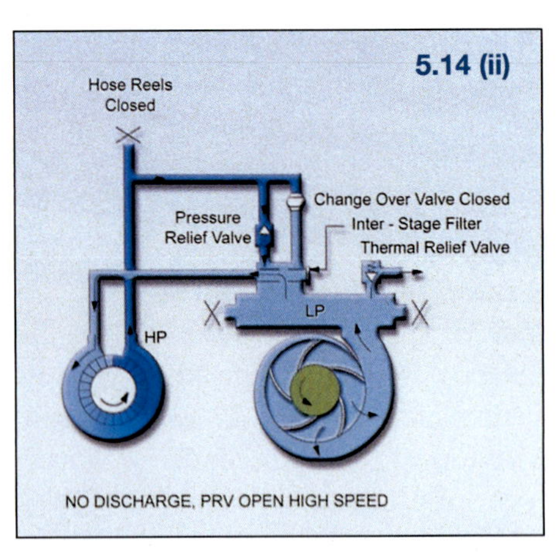

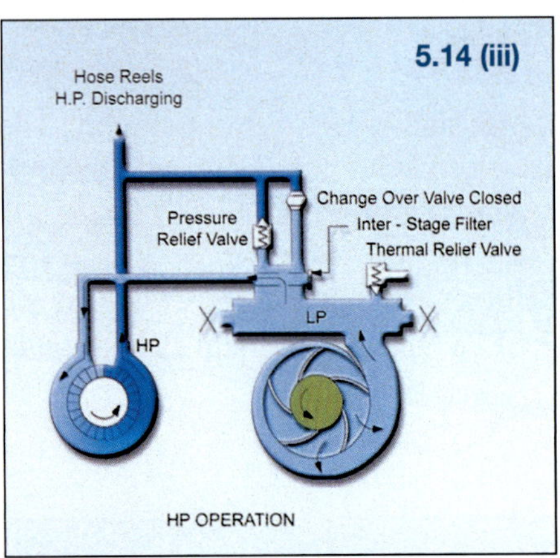

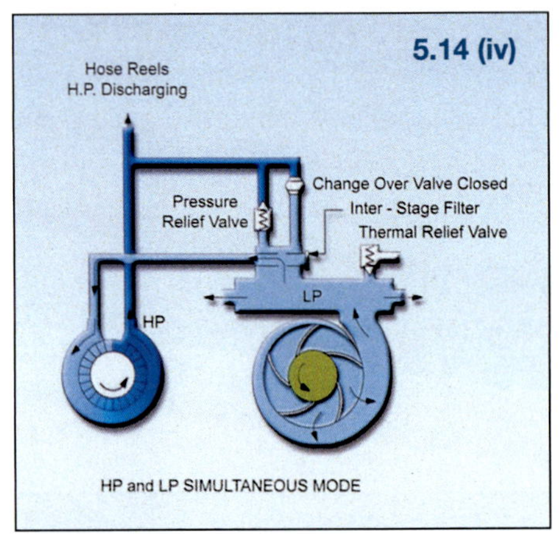

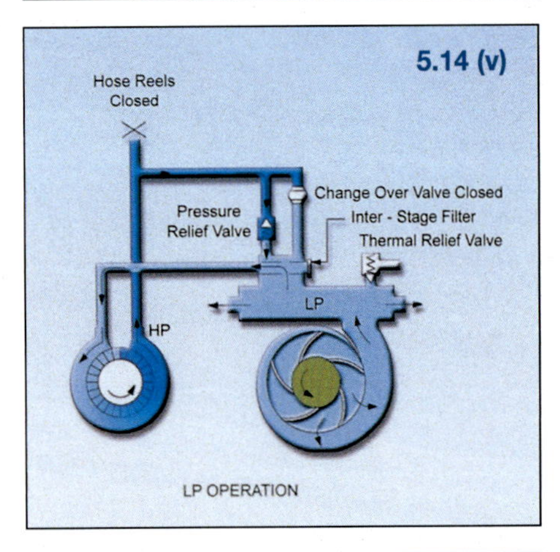

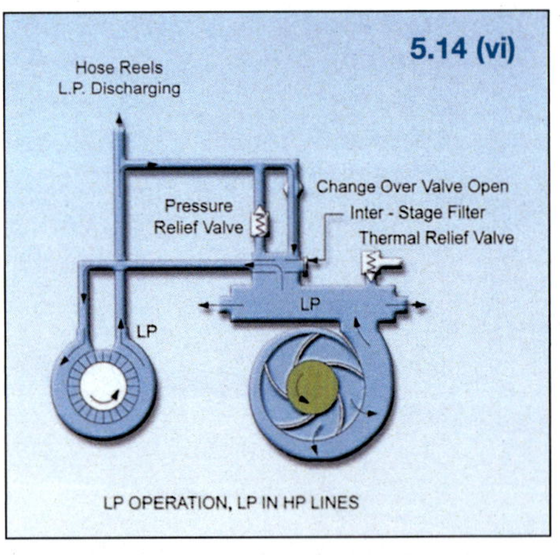

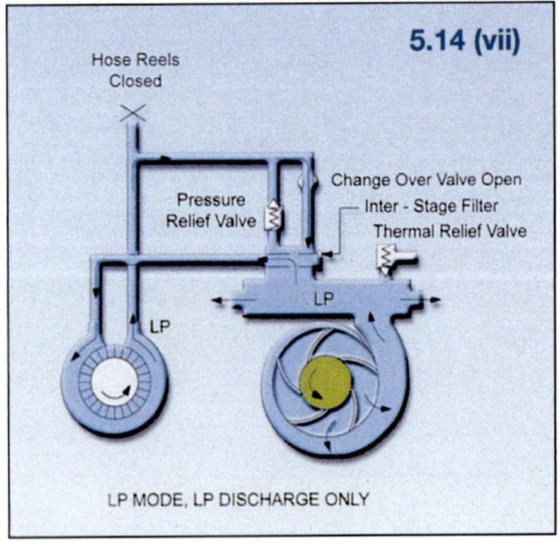

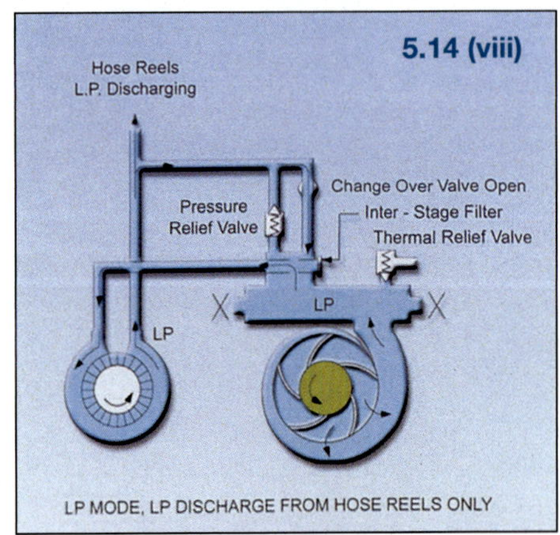

Figure 5.14 Modes of operation of the Godiva World Series Pump. (Courtesy of Hale Products Europe)

Figure 5.14 (v) shows LP operation but with the change over valve closed. The PRV opens to prevent excessive pressure in the HP stage.

Figure 5.14 (vi) shows both HP and LP deliveries and the change over valve open so giving LP operation in both lines.

Figure 5.14 (vii) shows the HP delivery closed and the LP delivery open. With the change over valve open, circulation from the HP stage is allowed so there will be no possibility of excessive temperatures.

Figure 5.14 (viii) shows the LP delivery closed but the HP delivery and the change over valve open. Under these circumstances only LP is available at the hosereels – a mode of operation suitable for damping down.

One disadvantage of the modern regenerative impeller is that the extremely close clearances involved in the high pressure stage require the water to be filtered before it enters that stage. The location of this inter-stage filter is shown in Figure 5.14.

Figure 5.15 shows a view of a typical pump bay with the change over valve indicated.

Figure 5.16 shows a diagrammatic representation of a multi-stage vehicle mounted pump of a type which is being used by a number of brigades. HP is achieved by the water passing successively through three additional conventional impellers.

Although, traditionally, the controls, instrumentation and deliveries for a vehicle mounted pump have been situated on, or close to, the pump itself, recent Health and Safety legislation has resulted in increasing concern about the noise level to which the pump operator is exposed. Because the solution of this problem, by the wearing of ear defenders, makes communication with other personnel difficult, some appliance manufacturers are

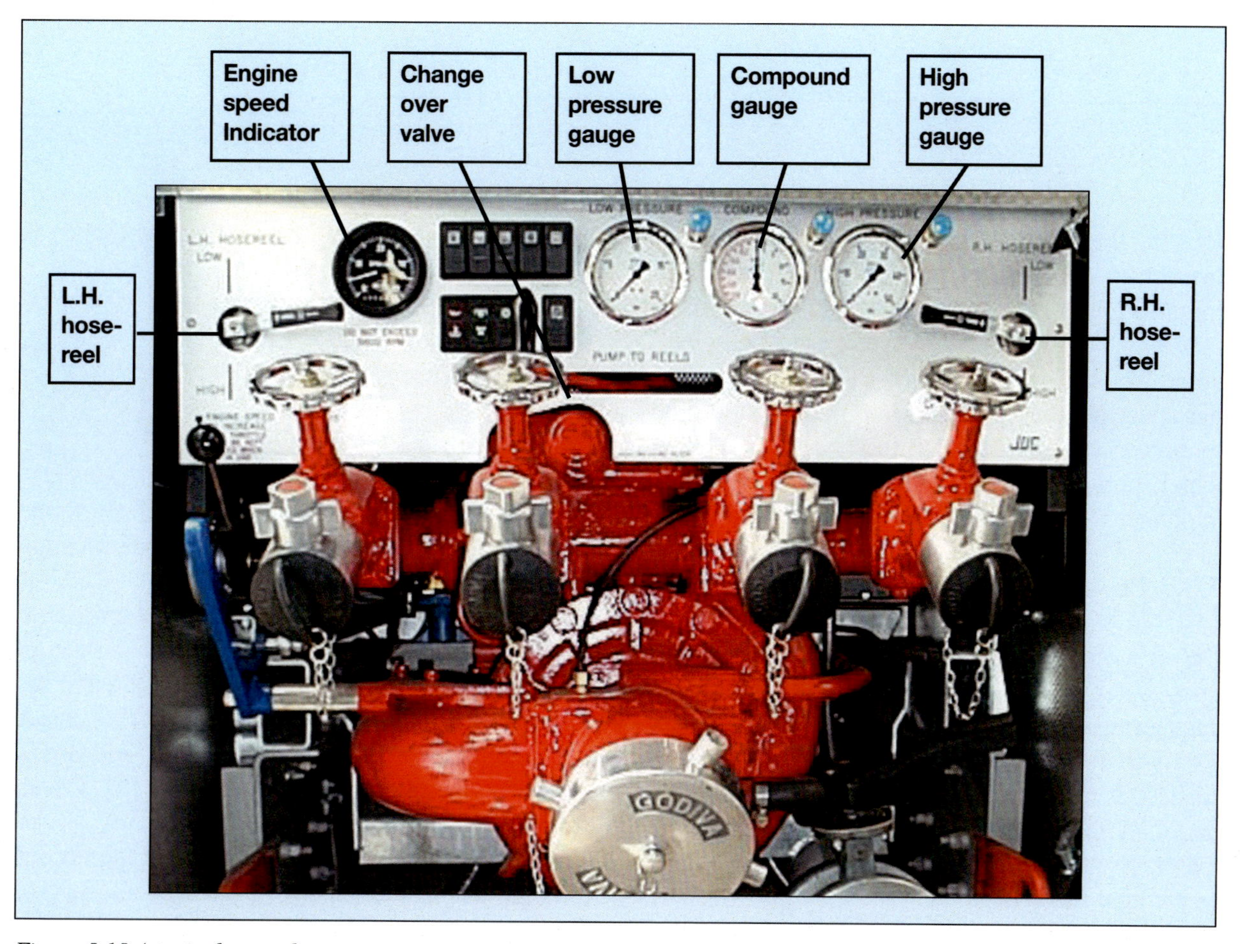

Figure 5.15 A typical pump bay. (Courtesy of Hale Products Europe)

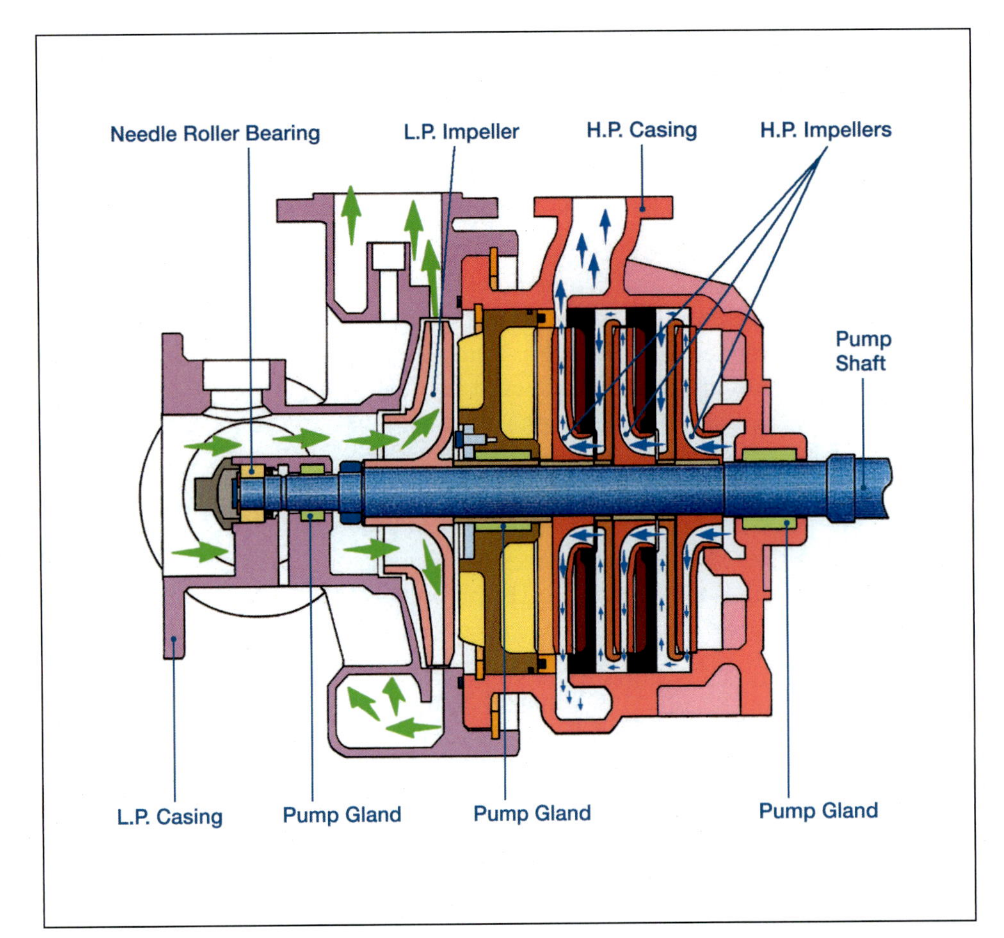

Figure 5.16 Diagrammatic representation of a multi-stage pump. *(Diagram courtesy of Rosenbauer)*

locating the pump in a soundproof booth with control levers etc. passing through rubber gaiters to minimise the amount of sound transmitted. Vehicle exhausts should obviously discharge at a point well away from the pump operator. At least one brigade has specified that its 4500 l/min pumps should be mounted centrally but with the operator's control panel, pump inlet and outlets at the rear of the appliance.

5.3.2 Primers for vehicle mounted pumps

Before a centrifugal pump can be got to work from open water, the air in the suction hose and pump casing must be expelled so that atmospheric pressure will force the water up into the pump. This process is called priming, and a device has to be provided for this purpose. It may be operated either manually or automatically, according to the type of centrifugal pump used.

The priming devices most commonly found on vehicle mounted pumps are:

- reciprocating;
- water ring.

(a) Reciprocating primers

The series of diagrams in Figure 5.17 show the principle of operation of a simple reciprocating piston primer. It consists of a small piston which is driven by an eccentric cam on the main pump drive shaft. (See also Figure 5.13.) On the induction stroke (Figure 5.17 (ii)), a vacuum is created in the priming valve body. Atmospheric pressure causes the automatic priming valve to open and air is drawn, by the piston, through the inlet flap valves from the suction tube. On the exhaust stroke (Figure 5.17 (iii)), air in the primer is forced out through the exhaust flap valves to atmosphere.

Each primer will discharge approximately 1 litre of water during a complete priming operation and a characteristic "popping" noise will be heard. Once pump pressure is generated it is communicated to the back of the piston which lifts off the cam so that the primer effectively disengages (Figure 5.17 (iv)).

Like a number of continental pumps, the Godiva World Series pump utilises twin horizontally opposed reciprocating primers.

The engine speed may be controlled automatically but, where it is not, the operator should take care not to exceed the speed recommended by the manufacturer – typically 2500 rpm.

(b) Water ring primers

A water ring primer is a form of positive displacement pump. It is widely used in the Fire Service, and is engaged and disengaged either manually or automatically.

The principle of operation is again very simple (see Figure 5.18). A vaned impeller with a hollow centre rotates in an elliptical housing around a stationary hollow boss which is a projection from the housing end cover. This boss has four ports in its periphery, which communicates with the primer suction and delivery connections.

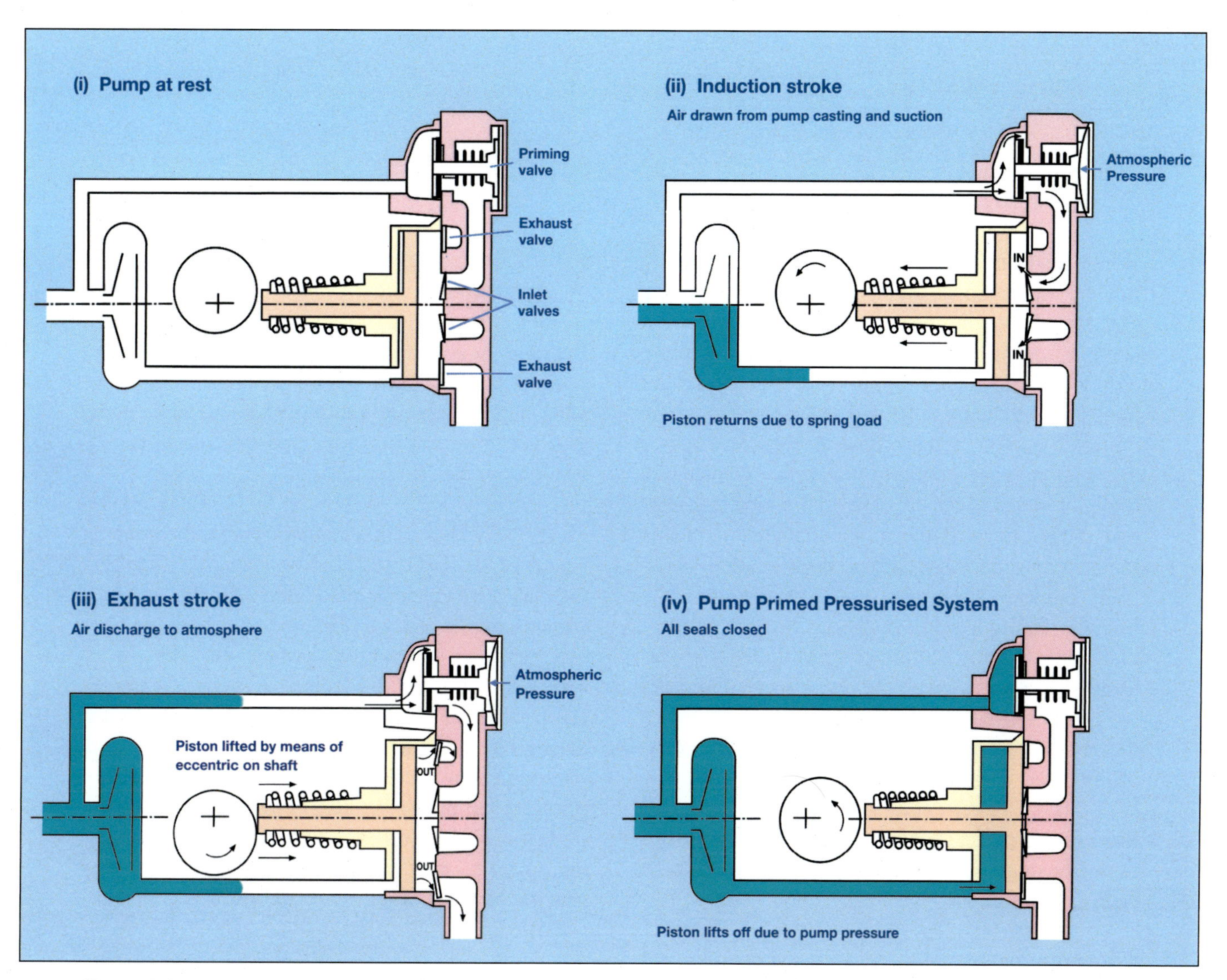

Figure 5.17 A reciprocating primer.

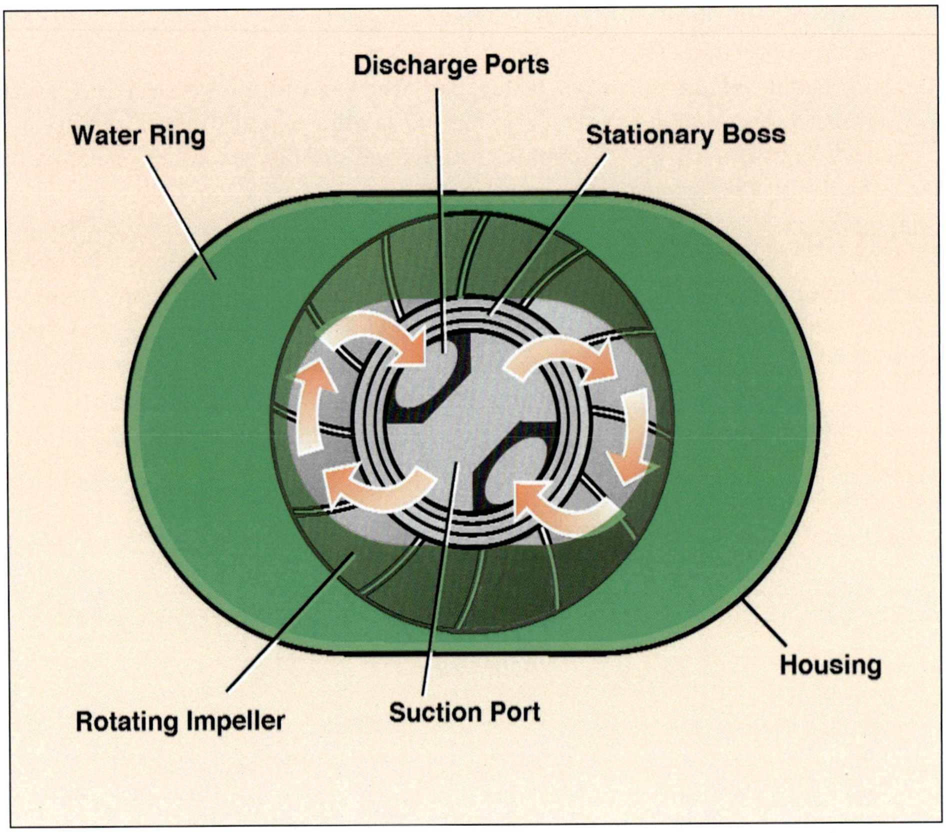

Figure 5.18
A water ring primer.

When priming commences and the impeller rotates, the liquid in the housing is compelled by centrifugal force to move outwards and follow the contour of the housing, thus forming a hollow elliptical vortex. This liquid "ring" rotates in the housing with the impeller and as it rotates from the minor diameter of the ellipse towards the major diameter it moves radially outwards between the impeller vanes. After it passes the major diameter and rotates towards the minor diameter it moves radially inwards. As the liquid moves radially outward between the vanes air is drawn into the impeller through ports in the centre which communicate with the suction ports in the central stationary boss and so with the pump suction line. As the liquid moves inward this air is forced through the impeller ports into the "discharge" in the central boss.

Since the impeller is located centrally in the elliptical housing there are two pumping actions in each revolution.

The appropriate engine speed for the operation of water ring primers is about 2500 rpm.

5.3.3 Cooling systems for vehicle mounted pumps

Because an appliance has to use its engine whilst stationary, its normal closed circuit cooling system, which also indirectly cools the automatic gearbox, is specially augmented by making use of water from the fire pump.

Figure 5.19 shows, in schematic form, the general layout of an appliance cooling system. A supplementary heat exchanger, adjacent to the gearbox coolant heat exchanger, is fitted to ensure the correct working temperature when the vehicle is stationary and the fire pump is in use. Water is taken from the delivery side of the pump, passed through the supplementary heat exchanger and returned to the suction side of the pump. This supply from the pump can also be used to cool the oil in the power take-off. The water is routed firstly to the power take-off and then to the supplementary heat exchanger before returning to the pump.

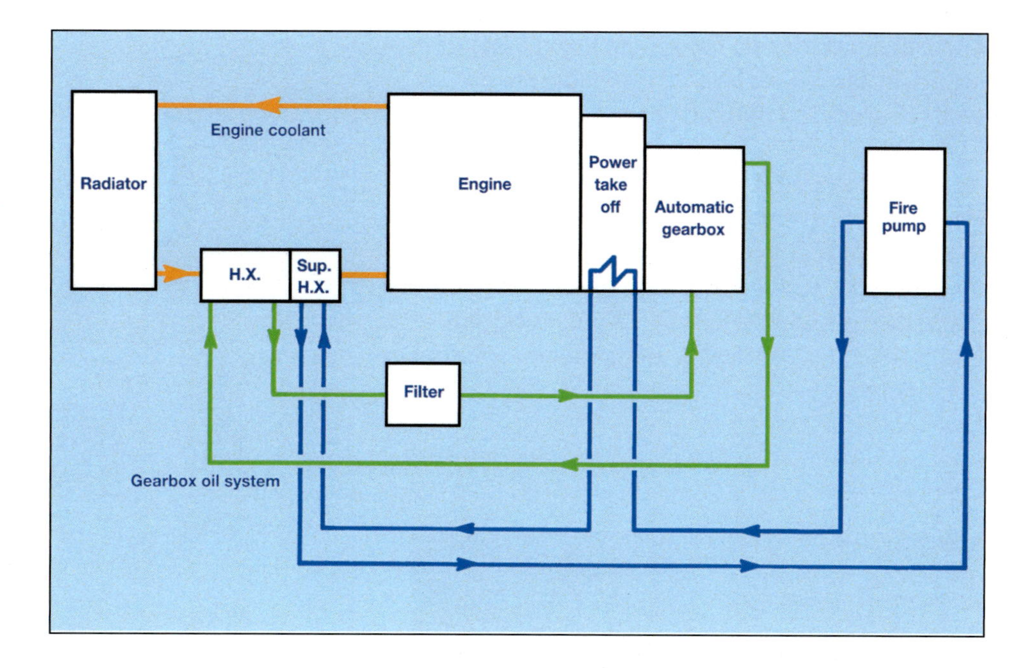

Figure 5.19 General layout of a closed circuit appliance cooling system. (Engine coolant: orange. Hydraulic oil: green. Fresh water: blue)

5.4 Portable pumps

5.4.1 Examples of portable pumps

(a) General

All brigades have, as part of their pumping capacity, pumps which can be manhandled into position. These are usually carried on appliances and are especially useful in areas where vehicles cannot get to water supplies. All have carrying frames and, depending on their weight, can be transported by two or four personnel. Most are driven by internal combustion engines but a few are electrically powered.

The current specification for portable pumps is that of JCDD/30 as amended in 1976. However a draft European Standard is in the course of preparation and eventually a new JCDD specification will be based upon it. JCDD/30 defines a lightweight portable pump (LWP) as a self-contained petrol-driven unit having a nominal output of at least:

Type 1 – 360 l/min at 5.5 bar, or
Type 2 – 1130 l/min at 7 bar.

It should be capable of being carried by two persons and not weigh more than 125kg. The specification also lays down an exhaust ejector priming system and stipulates a single-stage centrifugal pump, coupled direct to the engine, to keep weight down.

Although there are a number of pumps which come within this specification, manufacturers have tended to design and build to brigade requirements. This has led to anomalies, e.g. units called LWPs which are heavier than the specification and some that are lighter with, of course, performances to match. Figure 5.22 shows a variety of portable pumps with performances ranging from 250 l/min at 3.8 bar to booster pumps capable of delivering 2300 l/min at 7 bar.

(b) Electrically powered pumps

Electrically powered pumps used in the fire service will generally be of low capacity and are usually employed for pumping out where other types are not suitable: e.g. in basements where it is difficult to disperse exhaust fumes.

A typical example of a submersible pump used by a large county fire and rescue service is shown in Figures 5.20 and 5.21.

This pump weighs only 17kg and operates from a 110volt generator supply. It can pump from a submerged depth of 20m and provides a maximum output of 550 l/min. It can pump water down to a level of 2mm.

The pump can also be used to provide a firefighting jet in excess of 10m at maximum output for such purposes as firefighting at sea.

As with all electrically driven appliances involving water, electrical integrity is vital and firefighters should be careful to ensure that safety instructions for use and testing are carried out correctly. (Guidance is contained in DCOL 11/1988 – Fireground Electrical Equipment: Safety Requirements.)

Figure 5.20 (above) A firefighter carrying an electrically driven submersible pump. (Photograph courtesy of Grindex)

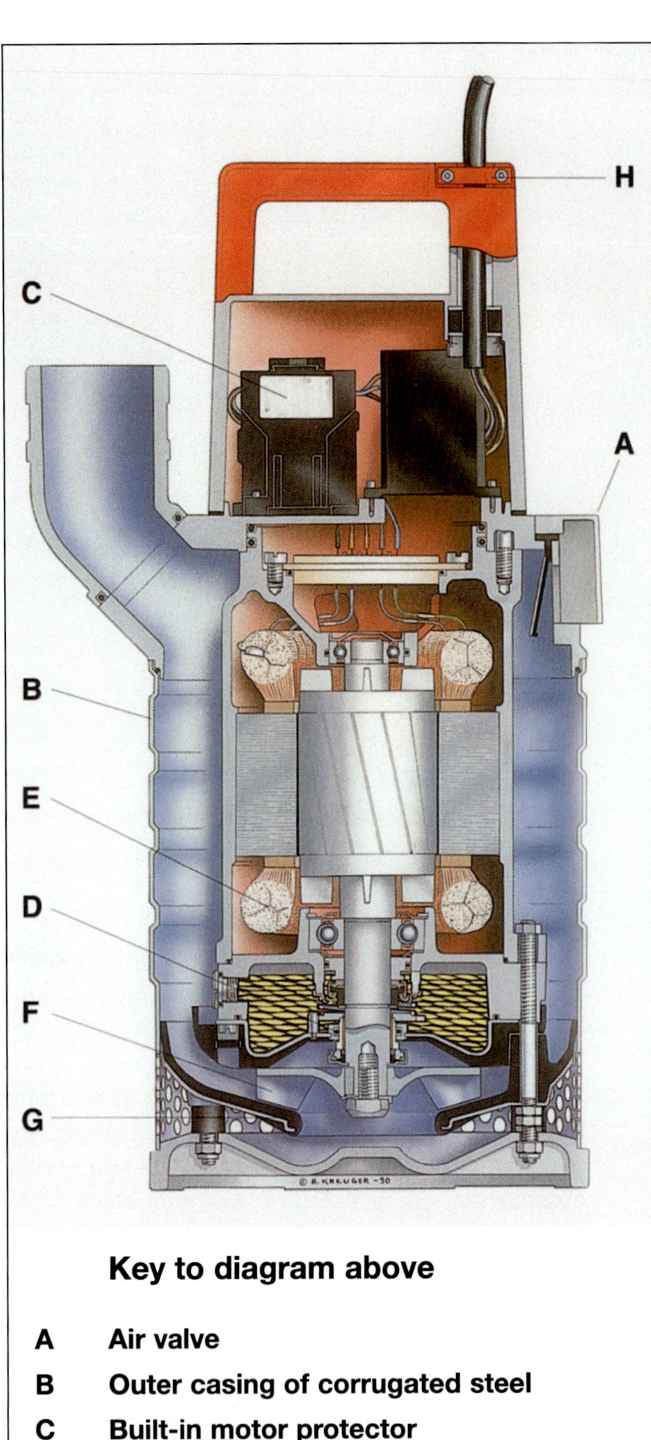

Key to diagram above

A	**Air valve**
B	**Outer casing of corrugated steel**
C	**Built-in motor protector**
D	**Patented shaft seal**
E	**Stator with class F insulation**
F	**Impeller of chromium-alloyed white cast iron**
G	**Adjustable diffuser**
H	**Clamped cable entry**

Figure 5.21 (right) A Grindex Minex submersible pump. (Diagram courtesy of Grindex)

5.4.2 Primers for portable pumps

Water ring and positive displacement primers, which have already been described in the section on vehicle mounted pumps, may also be found on portable pumps and an example of a small portable pump with a manually operated positive displacement primer is shown in Figure 5.22 ii. However, the majority of portable pumps employ a primer based on the ejector pump principle, described earlier in this Chapter, and use the exhaust gas from the engine as the propellant. Figure 5.23 illustrates the principle of operation.

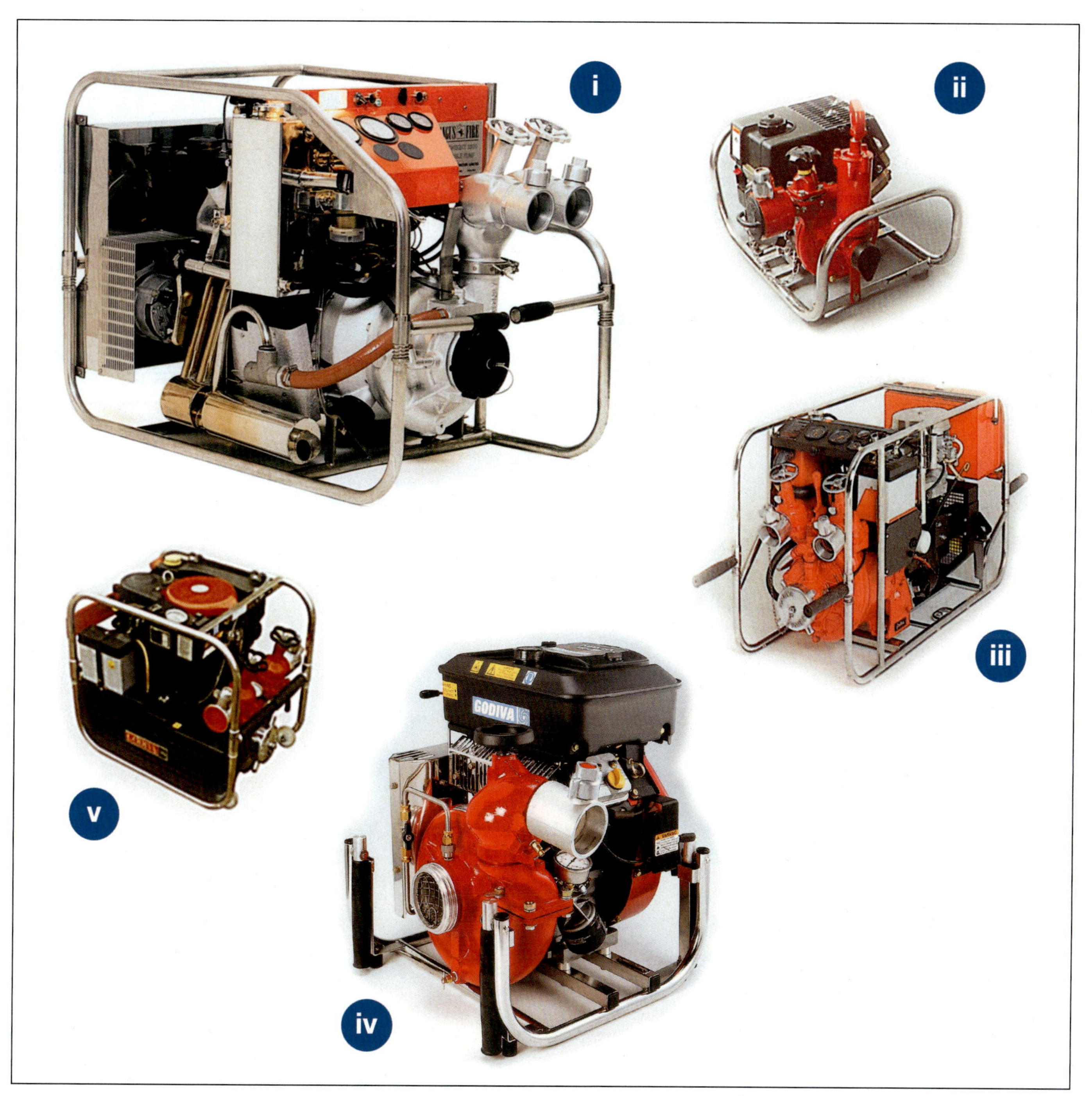

*Figure 5.22 A range of portable pumps (Clockwise starting top left: **i** Angus LW 2300; **ii** Godiva GP 250; **iii** Godiva GP 2300; **iv** Godiva GP 8/5; and **v** Godiva GP 10/10)*

With the engine running and the pump primed, the engine exhaust gases pass from the exhaust manifold, through the ejector housing and into the silencer in the normal way (Figure 5.23 top).

The ejector becomes operational when the primer rod (1) is pulled to the limit of its travel (Figure 5.23 bottom).

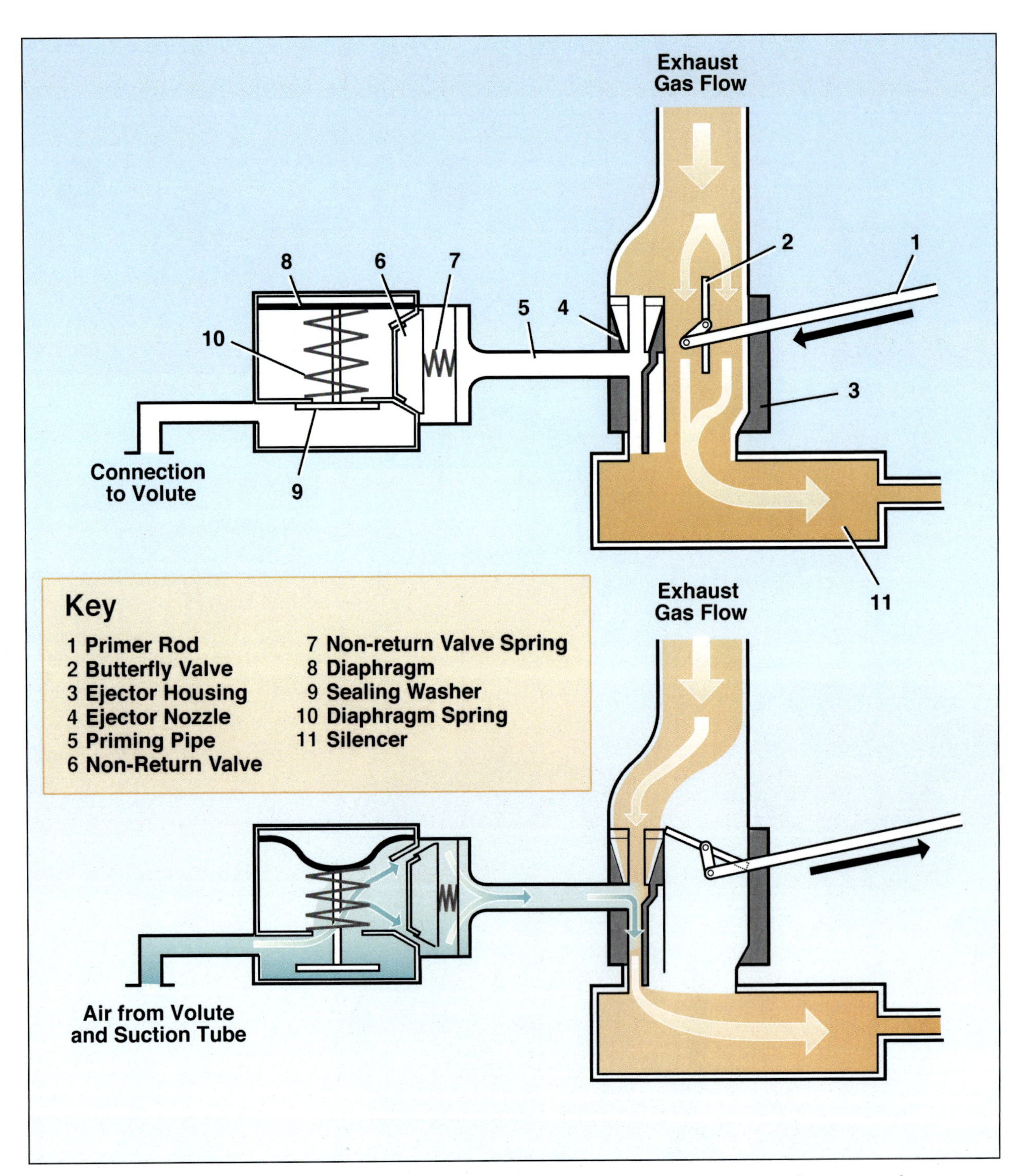

Figure 5.23 An exhaust gas ejector primer. Top: exhaust butterfly open and priming valve closed. Bottom: exhaust butterfly closed and priming valve open. The exhaust gases are shown in brown and the water in green.

The action of pulling the rod closes the butterfly valve (2) in the ejector housing (3) which then causes the exhaust to be deflected through the ejector nozzle (4) thereby creating a vacuum in the priming pipe (5) and, on the outlet side of the non-return valve (6). This causes the non-return valve to lift off its seating against the pressure of its spring (7). It also draws the diaphragm (8) down, which moves the sealing washer (9) against the pressure of the spring (10) into the open position. The depression then allows water to flow into the volute under atmospheric pressure.

As the pump is primed, evacuated air and then water pass through the system and into the silencer (11).

When pump priming is complete (indicated by the first positive needle movement of the pump pressure gauge) the priming rod (1) is released and returns to its static position under the influence of a return spring, causing the butterfly valve (2) to open. The exhaust gases then resume their normal path (Figure 5.23 top), the depression in the priming pipe (5) and outlet housing side of the non-return valve (6) is destroyed and the valve closes under the influence of its spring (7). The diaphragm (8) returns to its static position under the influence of the spring (10) and the sealing washer (9) contacts its seating.

Some modern pumps have an automatic priming valve so that only the exhaust butterfly needs to be closed to prime the pump.

With this type of primer, efficiency depends on the speed at which the gases leave the ejector nozzle. Priming is, therefore, carried out at high engine revolutions i.e. full throttle.

5.4.3 Cooling systems for portable pumps

Although many of the smaller portable pumps have an air-cooled engine, the majority employ an indirect closed water circuit system. In order to reduce weight and bulk, instead of a fan-cooled radiator a header tank is fitted within which are water cooling coils supplied from the delivery side of the pump (Figure 5.24). This water, after passing through the coils, is returned to the suction side of the pump. The tank itself contains water which

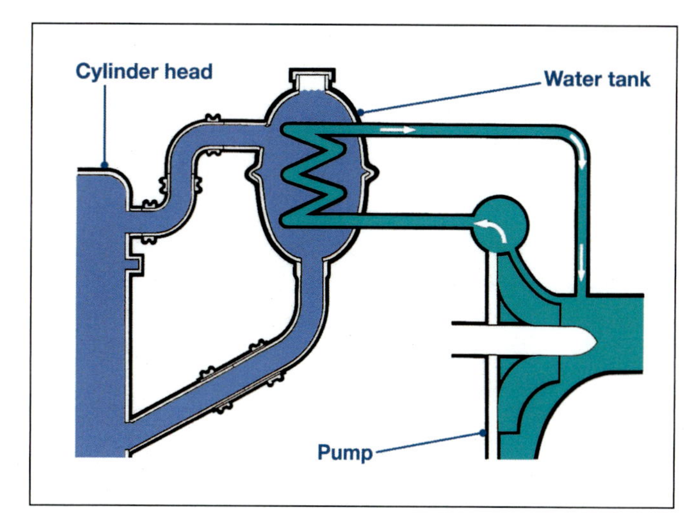

Figure 5.24 A closed circuit cooling system on a portable pump.

circulates round the engine block by means of a circulatory pump similar to that in an appliance, and can have anti-freeze mixture added to it.

It is important that, as soon as a portable pump of this type is started, there should be a supply of water to the pump to prevent overheating.

5.5 Safety

(a) As with all operational equipment, the purchase of all pumps, whether hand-operated, portable or vehicle-mounted, must be conducted so as to satisfy the requirements of the Provision and Use of Work Equipment Regulations 1992. These regulations require that all work equipment is assessed for suitability prior to purchase and is used, maintained, tested, inspected and disposed of safely.

(b) The major hazards associated with portable and vehicle-mounted pumps are:

- noise;
- high pressure.

In addition, portable pumps present significant risk to personnel from manual handling. (See also Fire Service Manual – Training.)

All these risks can only be effectively controlled by brigades having systems in place which ensure that:

- personnel at risk are aware of the hazards;
- equipment is purchased to minimise the risks;
- safe systems of work such as limiting exposure to noise are implemented;
- equipment is effectively maintained in a safe condition.

Hydraulics, Pumps and Water Supplies

Chapter

6

Chapter 6 – Pump Operation and the Distribution of Water on the Fireground

Introduction

This Chapter deals with the subject of practical pump operation and covers such points as getting to work from hydrants and open supplies, estimation of required pump pressures, fault finding, maintenance and testing and recent developments in automatic pump and tank fill controls.

6.1 Getting to Work from a Hydrant

Where only one pumping appliance is available, it is regarded as good practice to situate it as close as possible to the fire in order to facilitate good communications between the pump operator and other firefighters and to reduce friction loss in the delivery hose. **However, in situations where the running pressure in a main is low, some advantage may well be gained by positioning the pump close to the hydrant.** This is because the pressure in the main has to overcome friction, not only in the main itself, but also in the hydrant, standpipe and soft suction, and if the friction loss in the soft suction can be reduced, a greater quantity of water will be delivered before the supply is over-run. If the distance between the fireground and the hydrant is more than a few hose lengths it may be necessary to set up a relay (see Chapter 7) in order to obtain the maximum flowrate from the main.

After the standpipe (preferably double-headed) has been attached, the hydrant should be 'flushed' i.e. turned on briefly to expel any foreign matter from the outlet. The pump collecting head should then be connected to the standpipe by soft suction, usually standard 70mm hose. It is desirable to lay twin lines of hose at the outset, to ensure that the maximum output of the main is available to the pump. The strainer should be left in the suction inlet in order to protect the pump from stones etc. which occasionally find their way into water mains. As the supply is pressure-fed the pump does not require priming and the primer should never be used.

As soon as the pump has been connected, a delivery valve should be opened and the hydrant turned on again, to allow air to be expelled from the hose and pump and avoid the build-up of excessive internal pressure. When water begins to flow from the valve it can be shut down and the delivery hose connected.

When all connections have been made and the delivery valve is opened again, the compound gauge and pressure gauge will indicate the pressure of the supply to and from the pump. If pump revolutions are increased to raise the delivery pressure, the reading on the pressure gauge will rise and that on the compound gauge will fall (because of increased pressure losses in the main and soft suction).

Depending on the main's capacity, a point may eventually be reached when the pump is delivering all the water the hydrant can supply. The compound gauge will then read zero and any further attempt to increase the output from the pump, e.g. to supply more jets, by increasing its speed will over-run the supply, tend to create a vacuum in the soft suction and so cause it to collapse under atmospheric pressure. As explained in Chapter 4, taking water from a second, nearby hydrant on the same main may prove to be of only marginal benefit.

A competent pump operator will be able to tell, by feeling the soft suction, whether there is sufficient water for additional jets, in which case the hose will be quite hard, or, if it feels soft, that the limit of the main's supply capability is being reached.

There are several rules on the use of hydrants that firefighters should bear in mind:

- **The hydrant valve should be opened slowly to allow the hose to take up the pressure and expel the air.**

- **The valve should be closed slowly to prevent water hammer.**

- **The valve should not be opened if the pit is flooded unless a standpipe or hose is first connected. Otherwise, if the main is empty, water from the pit could enter and pollute the supply.**

- **If a hydrant has no water available e.g. it has burst or the supply has been shut down, firefighters should ensure that the valve is shut before unshipping the standpipe or hose.**

- **If possible firefighters should avoid collecting water from a street water-main and a dirty-water supply simultaneously. Although collecting heads have non-return valves any defects could cause the street main to become polluted.**

- **The hydrant valve should be properly closed to avoid leaks and any frost valve checked to see that it operates correctly. The pit should be left clear of water and debris and, if available, the outlet cap replaced.**

If the appliance has automatic tank fill control (see Section 6.8), it is normal practice to connect the soft suction to the tank fill inlets so that water is supplied to the pump via the tank. This has the advantages of:

(1) enabling pumping to continue without the interruption which might be caused by switching over from the tank supply to the pressure fed supply;

(2) allowing the pump operator greater control over delivery pressures which are no longer affected by pressure fluctuations in the main supply.

When the demand for water is greater than the ability of the hydrant to supply it, the operator will, of course, find it impossible to maintain the water level in the tank.

6.2 Getting to work from open water

Personnel who are at risk of accidentally entering water should be adequately protected. Protection can range from a life-jacket to, in particularly hazardous circumstances, lines and harnesses.

6.2.1 Setting up

The vacuum needed to lift from open water means that only hard suction, specially designed to with-stand external pressure, can be used. This type of hose is much less flexible than delivery hose and must be laid out carefully to avoid acute bends.

The pump should be positioned as close to the open water supply as possible with the suction laid out in a straight line and coupled up using suction wrenches to obtain airtight joints. A metal strainer should be placed on the end of the suction hose and, if it is considered necessary, a basket strainer over it.

Before the suction is lowered into the water it should be secured by a line to the appliance or other suitable point, to take the greater part of the weight off the inlet coupling and also to make it easier to draw up the hose to clean the strainer. If the suction is passing over a rough edge e.g. wall, quayside, dam, the underside should be protected from chafing. Care must be taken to avoid any sharp vertical bends above the suction inlet, as these could lead to air pockets causing a poor supply (Figure 6.1).

To get the most satisfactory results at the pump, the top of the strainer should be submerged to a depth of at least three times the suction hose diam-eter. At any less distance the strainer will tend to rise, causing vortices to form (Figure 6.2) and air

Figure 6.1 The right and wrong methods of leading suction hose over a wall or other obstruction.

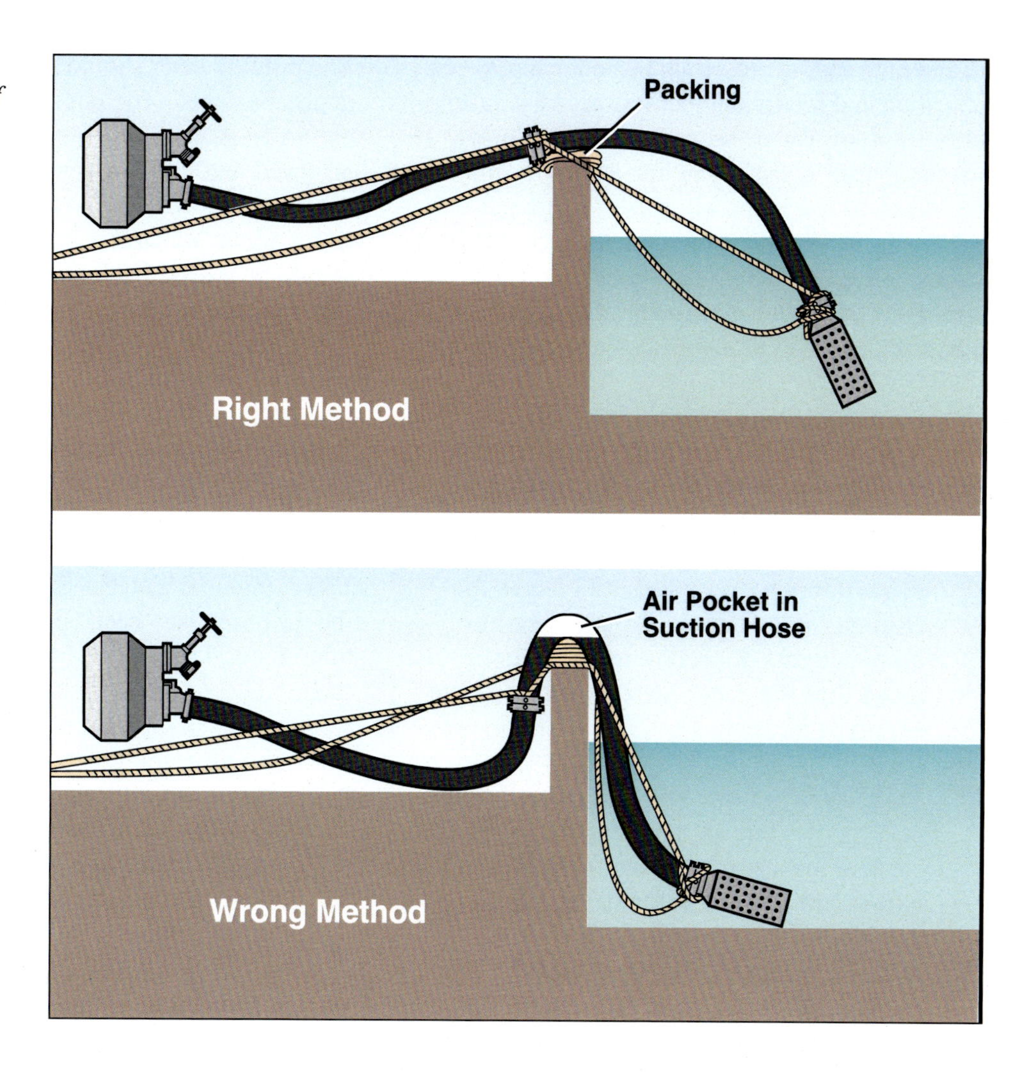

Figure 6.2 Sketch showing vortex formed through allowing the suction strainer too near the surface of the supply.

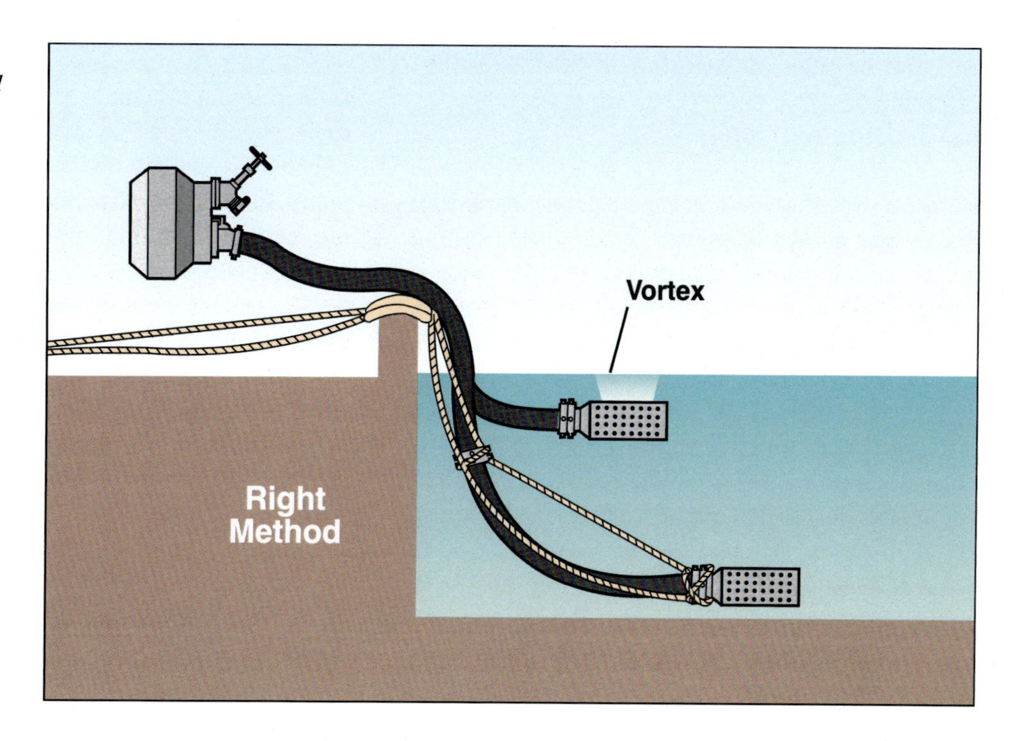

to enter the suction, resulting in a poor supply with crackling and inefficient jets and, probably, a complete loss of water necessitating re-priming. If working from a fairly shallow supply, it may be necessary to use a special low-level strainer or to anchor the strainer to the bottom using a weight. A strong current in a river may also tend to make suction hose rise and it should, if possible, face upstream, and will probably also need anchoring.

When getting to work from an open supply:

- **Situate the pump as close as reasonably possible to the supply in order to reduce the vertical lift and the number of lengths of suction hose required (and hence the friction loss). However, in deciding how many lengths of suction to use, bear in mind that the water level may drop if the supply is limited.**

- **Make sure that the strainer is free from obstruction but not too near the water surface (otherwise vortices may form and so allow air to enter the stream).**

- **When a substantial vertical lift is unavoidable pump operators should be aware that the pump's performance will be significantly reduced and that increasing the throttle setting beyond a certain point will actually reduce the output because of cavitation in the impeller.**

6.2.2 Using a primer

Before a centrifugal pump can be got to work from open water it must be primed. To do this a priming device, either manually operated or automatic, is brought into action. The various types of primers are described in Chapter 5.

The throttle should be set at the recommended priming speed; with reciprocating primers this is normally about 2500 rpm, and may be controlled automatically. Water ring primers also require about 2500rpm. Pumps fitted with exhaust ejector primers, however, need to be run at full throttle to prime. On some portable pumps the throttle setting is automatically adjusted until priming is completed.

The priming lever should then be operated and the gauges watched. The compound gauge will register an increasing vacuum reading as the lift is achieved and, when a constant vacuum reading is obtained, the pressure gauge should show a positive reading. When this happens the priming lever should be released and the throttle eased back. On some pumps the primer is automatically engaged when the pump is started and automatically disengaged when priming is complete. With an exhaust gas ejector primer it may be necessary to crack a delivery valve to allow the escape of air from the pump casing before releasing the primer and throttling back.

The compound gauge will indicate whether or not the priming has been successfully effected. If, after about 45 seconds, lift has not been achieved then:

(a) the lift is too great, or
(b) there is an air leak on the suction side of the pump or in the pump itself (glands, gauge connections etc.), or
(c) one of the defects mentioned in Section 6.6 below applies.

Before attempting to prime again, all joints and connections should be checked for air leaks. If priming still cannot be achieved, it may be necessary to try to re-position the pump in order to reduce the lift.

If the priming system in a pump is defective and no replacement pump is available, priming can be effected manually with the help of a blank cap. The pump should be connected to the suction hose in the normal way, but with a delivery valve left slightly open. The last length of suction hose, with no strainer on the end, should then be held above the level of the pump, and water poured in until it fills the pump casing; this will be indicated by water flowing from the open delivery, and the valve should then be closed. More water should be poured into the suction hose until it is completely full, the end being held vertical for this purpose. While it is being filled, the hose should be moved about in order to release any air pockets. The blank cap should then be lightly screwed onto the end of the suction.

The suction should then be lowered into the water, with a light line such as a belt line bent onto the blank cap to prevent its being lost. The pump should be run at a fairly brisk speed and the blank cap removed under water, either by hand or with a tool. If all joints are airtight and no air pockets are left, the pump will now lift water without difficulty. The suction strainer(s) should then be attached, provided that this can be done whilst keeping the end of the suction adequately submerged.

An alternative method of dealing with the problem, provided a second pump is temporarily available, is to fit a collecting head to the outlet of the faulty pump and then to connect it, via hard suction, to the inlet of the second pump, which may then be used to do the priming. This method has the advantage that it is unnecessary to disturb the suction, strainer etc. already connected to the faulty pump.

6.2.3 Holding water

When priming has been effected, the water should be held until the branch operators are ready to receive it. Sufficient impeller speed should be maintained to hold the water at a pressure of not less than 1.5 bar. This will ensure circulation of sufficient water through the cooling system. The pump operator must, however, remember that the length of time for which the pump may be run under these conditions will be limited unless a thermal relief valve is fitted. This is because the impeller imparts energy to the water which is converted to heat, and, with the added heating effect of the engine, the temperature in the pump casing will rise, causing possible overheating problems. The functioning of a thermal relief valve, which is fitted as standard to many modern pumps, is described in 5.3.1 When the order 'Water on' is received, the delivery valve should be opened carefully and pump revolutions gradually increased to give the required pressure at the branch without a sudden branch reaction.

6.3 Cooling systems

As stated in Chapter 5, all modern vehicle mounted pump engines and the larger portable pumps have closed circuit cooling systems, augmented by secondary coolant supplied from the pump itself, and a gauge to tell the pump operator the

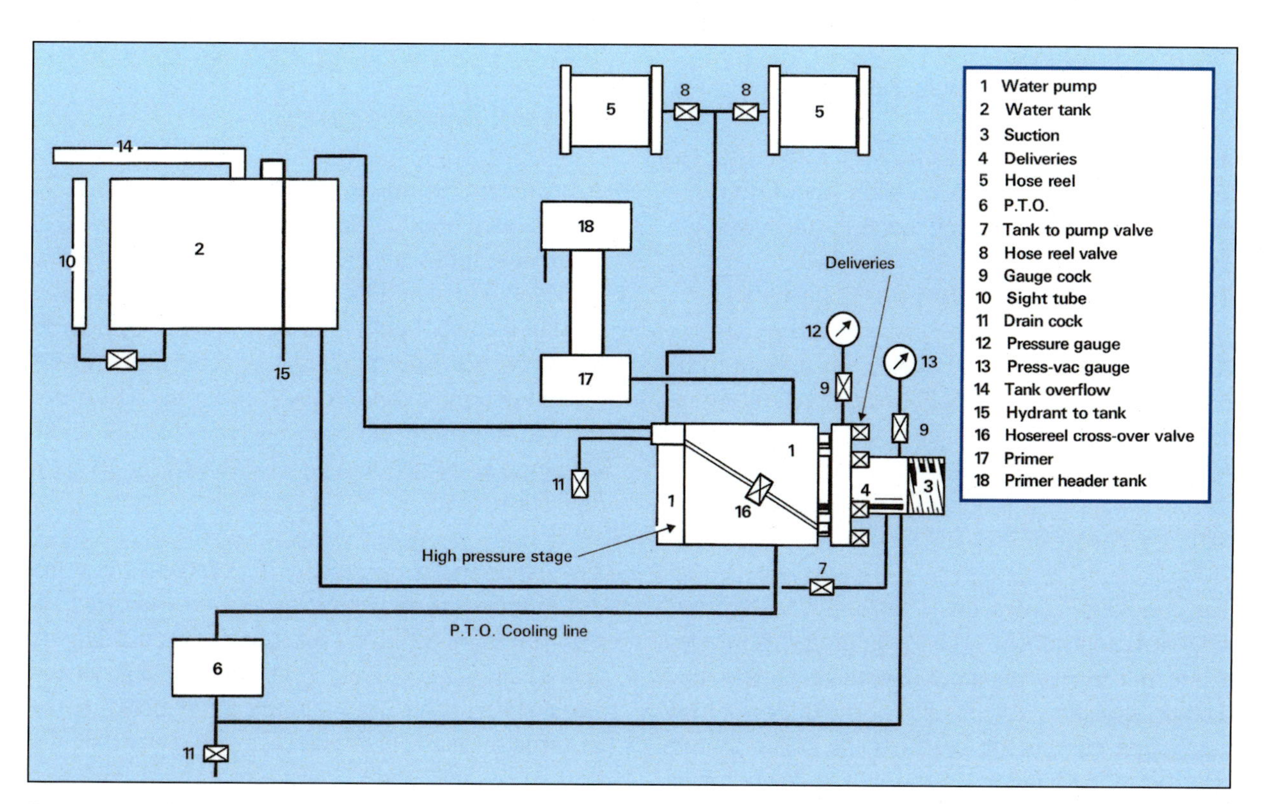

Figure 6.3 Schematic layout of a typical appliance pipework system.

temperature of the engine coolant (see 6.4.6 below). Where a water-cooled portable pump does not have a temperature gauge, the operator must keep watch on the filler cap of the header tank to see that the water does not boil. If it does, the pump should be shut down immediately, allowed to cool, and the tank then topped up with fresh water if necessary. Some models incorporate an automatic device to shut down the engine or reduce its speed if overheating occurs.

Operators should ensure that they keep an adequate flow of water through the pump as per the manufacturer's recommendations.

6.4 Instrumentation

The number of gauges on modern appliances has proliferated but, for the pump operator, there are usually seven that are essential:

- Pressure gauge(s)
- Water tank contents
- Compound gauge
- Oil pressure
- Tachometer (rpm)
- Fuel tank contents
- Engine coolant temperature

Most of these can be seen in Figure 5.15.

In recent years, some brigades have also requested the fitting of flowmeters on each pump delivery, but appliances so equipped are still in a minority.

6.4.1 Pressure and compound gauges

A pump operator should watch the pressure and compound gauges carefully whilst operating as they give a reliable indication of how the pump is performing and how the water supply is being maintained. The working of these gauges is dealt with in Chapter 2.

The compound gauge, on its vacuum side, registers any variation of lift when pumping from open water, but pump operators should remember that the vacuum shown on the gauge includes not only the height of the lift but also the other factors referred to in Chapter 3, Section 3.2. For practical purposes, -0.1 bar would indicate a lift of about one metre but a pump lifting approx 3 metres may show a reading of -0.4 bar. The extra -0.1 bar represents the losses due to the other factors.

6.4.2 Tachometers

It is helpful to a pump operator to know the speed at which the engine is running. The number of rpm can be critical when priming with a reciprocating primer and also it is a very good indication of the efficiency of a pump under test. When new, a pump is capable of delivering a certain quantity of water at a given pressure, suction lift and pump speed. If, later, a higher pump speed is required to obtain the same output it is an indication that the efficiency has dropped.

6.4.3 Water contents gauge

On many appliances the depth of water in the water tank is shown by a vertical calibrated transparent pipe. When filling the tank, from whatever source, the pump operator should watch the gauge to ensure that the tank is not overfilled. If an appliance is standing on soft ground, any overspill could possibly cause the appliance to become bogged down, and in a water 'shuttle' situation (see Chapter 7) this could have serious implications.

6.4.4 Oil pressure gauge

All modern pumping appliances and portable pumps are fitted with oil pressure gauges. On appliances these are usually in the cab and any low pressure may in addition be indicated by a warning light in the cab. Pump operators must remember to check the pressure regularly, especially under protracted pumping conditions. Neither the gauge nor the warning light indicates the level of the oil in the sump and this must be ascertained by taking a dipstick reading.

The experienced operator will come to know the different pressures which should be indicated on the gauge for different conditions. A cold engine should give a relatively high reading but, as the engine warms up, this should drop back to the working pressure. This pressure will depend on the type of engine, and a worn engine will probably give a lower reading than one in good condition. If the pressure drops back appreciably the engine should be stopped immediately, but the operator must remember the firefighters operating branches and try to make alternative arrangements for their supply.

6.4.5 Fuel tank contents

This gauge is also found usually in the cab. Provided that the driver/pump operator has completed the necessary checks on taking over the appliance there will be an adequate supply of fuel at the commencement of pumping, but allowance should be made for the distance run to the incident. The operator should have a reasonable idea of the consumption when pumping, but should still check the gauge at frequent intervals and remember to inform the officer-in-charge, in good time, of the need to obtain further supplies.

6.4.6 Engine coolant temperature

This gauge is found either on the instrument panel in the cab or by the pump controls. If it indicates an excessive rise in temperature, the engine should be shut down immediately.

6.4.7 Flowmeters

Whilst the delivery pressures at pumps need to be constantly monitored for safety reasons, there are a number of practical advantages to the measurement of flowrates from pumps and in individual hoselines.

Following trials conducted by the Fire Experimental Unit (FEU) in three brigades during 1989 it was concluded that flowmeters, fitted to the main deliveries and hosereel supplies of fire service pumps, would be of benefit in the following ways:

(i) **Flowmeters can ensure that flowrates at the branch can be handled safely, and reduce the need to calculate the pressure loss between the delivery and the branch. Also the pump operator can quickly identify burst lengths, because of the unexpectedly high flow, and vandalised dry risers if there is flow after the riser has been charged.**

(ii) **It is no longer necessary for every delivery to work at the same pressure. Different size nozzles can be controlled by the pump operator and, when operating an aerial appliance, the correct flowrate for the monitor can be maintained, when its height is changed, simply by adjusting the pump rpm.**

(iii) **By assisting the pump operator to identify the causes of changes in the readings of the pump pressure gauges.**

(iv) **Where hosereel induction systems are fitted, which induce additives into the pump, the use of hosereel flowmeters enables the pump operator to supply the correct quantity of water to one or two branchmen or, if no branches are in use, to turn off the supply.**

(v) **Experience indicates that, with flow meters, operators generally run pumps at reduced rpm settings, whilst still providing effective jets or spray for firefighting, and so reduce wear and tear on the appliance.**

(vi) **The control and monitoring of available water supplies and of the total amount of water used at incidents (e.g. ship fires) is greatly improved.**

(vii) **Flowmeters have proved useful in the evaluation of foam monitors and for special service calls where there is a requirement for the delivery of a specified amount of water (e.g. chemical incidents where it may be necessary to dilute spillages by a known amount).**

In spite of these positive conclusions, flowmeters are being included in specifications for new appliances by only a limited number of brigades at the present time. Figure 2.14 shows a pump fitted with analogue flowmeters on each of its main deliveries.

At the time the FEU report was written, most flowmeters were of the paddle-wheel type, but lately the electromagnetic type seems to be preferred. Both are described in Chapter 2.

To achieve sufficient accuracy for fireground use (say + or -10%) careful design of the installation in the deliveries and calibration are required.

6.5 Estimation of required pump pressures

Pump performance, which is discussed in Chapter 5, is usually rated in litres per minute at a pressure of 7 bar with a stated suction lift of 3m and an experienced pump operator will know the approximate quantities his pump will deliver at different pressures and from different suction lifts. The operator should also know the approximate rates of discharge from the types of nozzles used by the brigade and, where it is significant, the approximate friction loss in the type of hose used.

For safety reasons, when the pump operator first supplies water it should be to give moderate nozzle pressure. If a request is received for more, or less, pressure, it should normally be altered by 1 bar at a time. If a number of branches are at work at different levels, the operating pressure may have to be a compromise as each line, ideally, would be supplied at a different pressure. In extreme cases the correct pump operating pressure for one line might be such as to endanger a firefighter operating from a different line and, in such cases, it may be necessary to shut down one or more branches and reconnect them to another pump.

To estimate the required pump pressure, 0.1 bar for every metre the branch is above the pump should be added to the branch operating pressure and the same amount deducted for every metre it is below the pump. The friction loss in the hose, which is now of greater significance because of the trend to the use of smaller diameter hose on the fireground, should also be added. Normally this estimation of pump pressure should be the task of the branch operator who is better positioned to assess these factors than the pump operator.

Friction loss calculations (Chapter 1), for the hose diameters most likely to be used on the fireground, give the values in Table 6.1 for the flowrates indicated.

Example 1

What pump pressure is required to supply a branch which delivers 400 litre/min at 5 bar when it is working on the fifth floor of a building, 15 metres up, using six lengths of 64mm hose?

Pressure required at the nozzle	5.0 bar
Pressure equivalent of 15 metres head	1.5 bar
Pressure loss in 6 lengths at 0.2 bar/length	1.2 bar
Pump pressure required	**7.7 bar**

Had 45mm hose been used, the friction loss alone would have been 6 (6 × 1.0) bar and the required pump pressure no less than 12.5 bar to achieve the same discharge! In practice a much lower operating pressure would probably be requested with the result that the discharge from the jet would be less than desired. However, the penalty, in terms of flowrate, for a reduced pumping pressure is not as severe as might be expected. Ignoring the height

Table 6.1 ***Friction loss per 25m length of hose for various flowrates and hose diameters***

diameter (mm)	friction loss (bar) for the flowrates (litre/min) indicated									
	200	400	600	800	1000	1200	1400	1600	1800	2000
38	0.7	2.9	6.8							
45	0.3	1.0	2.1	4.0	6.0					
64		0.2	0.4	0.7	1.0	1.2	2.0	2.7	3.3	4.2
70		0.1	0.3	0.4	0.7	1.0	1.3	1.7	2.2	2.7
90			0.1	0.2	0.3	0.4	0.5	0.7	0.9	1.1

Impractical

Consider

Negligible

Note: If lines are twinned, the loss in each line will be reduced to a quarter of the value in the table.

factor, if the pump pressure is reduced by 50%, the flowrate will decrease by approximately 30%.

Whilst it is unrealistic to expect precise calculations of friction loss on the fireground, operators should be aware of the magnitude of the losses which might be incurred, particularly with the smaller hose diameters. When policy decisions are being made concerning the types of nozzles and hose diameters to be carried on appliances, care should be taken to ensure that the two are compatible and do not result in a need for unreasonable pumping pressures.

It is flowrate which extinguishes fires – not pressure

For 19mm hosereels, with all the hose laid on the ground, a typical flowrate of 100 l/min results in a pressure loss of approximately 10 bar over 55 metres. This increases to approximately 16 bar with most of the hose wound on the drum.

6.6 Identification of faults

The ability to interpret, intelligently, what the gauges are indicating is the hallmark of a good pump operator. It should be possible not only to recognise a developing situation but also to tell, within fairly close limits, the cause. The following checklist covers the most common faults. The methods of rectifying them are shown only where they are not self-evident.

6.6.1 Working from a pressure-fed supply

(a) Failure or reduction of water supply

This may be caused by:

(i) Failure of the supply e.g. fractured main or burst length of hose between supply and pump.
(ii) Choked internal strainer of the pump.
(iii) Over-running the supply.

(b) Increased delivery pressure whilst at work

(i) The closing down of a hand-controlled branch.
(ii) Debris fallen onto the delivery hose or a vehicle's wheels parked on it.
(iii) A bad kink in the delivery hose
(iv) A remote possibility, when using small diameter nozzles, that a stone may have passed through the internal strainer, pump and hose and blocked a nozzle.

(c) Decreased delivery pressure whilst at work

(i) A burst length of hose on the delivery side of the pump.
(ii) A hand-controlled branch being opened up.

6.6.2 Working from open water

(a) The pump fails to prime

If the pump fails to prime, the compound gauge will show either no vacuum reading or a very high vacuum reading:

No vacuum reading
(i) Suction strainer not adequately submerged (see Figure 6.2).
(ii) Faulty joints on suction hose and inlets or air leaks in suction hose.
(iii) An open drain cock, or loose drain plug, in the pump casing or air leak in a gauge connection.
(iv) Delivery valve not seating properly. The non-return valves should seat and hold a vacuum.
(v) Defective pump seal – should become apparent through regular testing before it prevents priming.
(vi) Incorrect seating of the exhaust valve in an exhaust gas ejector primer.
(vii) Defective drive to mechanically driven primer.
(viii) Defective priming lever linkage.
(ix) Lack of water in water-ring primer.
(x) Compound gauge cock closed.

Many of the above faults should become apparent during regular testing (see Section 6.7).

A very high vacuum reading
(i) Blocked metal or basket strainer.
(ii) Collapsed internal lining of suction hose.

(b) Changes in the compound gauge reading

Some changes in the compound gauge reading, whilst the pump is at work, are to be expected. They will result from such factors as:

(i) changes in the level of the open water supply;
(ii) changes in the demand from the pump. A greater demand will mean increased losses on the suction side of the pump with a consequent reduction in the gauge reading.

(c) Cavitation

When pumping from an open supply, the pressure on the inlet side of the pump can be so low that the water can vaporise at the ambient temperature. Vapour bubbles then form (this is known as cavitation) and the pump gives off a distinctive rattling sound. When this sound is heard the pump operator should appreciate that the maximum flowrate for the prevailing conditions has been achieved and should throttle back until the sound disappears. The compound gauge will not normally give an early indication that cavitation conditions have been reached. The danger of cavitation is obviously greater if the water is hot (see Chapter 3).

(d) Crackling jets

A crackling jet is caused when air is taken into the pump with the water. The water and air are pressurised and the air expands explosively as it leaves the nozzle. Again, the explanation may be that the strainer is too near the surface, allowing air to be drawn down, or there could be a slight leak on the suction side of the pump.

(e) Mechanical defects

Regular servicing and testing should reduce mechanical defects to a minimum and major problems should not, therefore, occur on the fireground. However firefighters should be aware that pumps are usually designed with a number of drillings, drains or "holes to atmosphere", blockages of which may seriously affect the performance. Drips from practically any of them are indicative of leaking diaphragms, faulty drain valves etc. and, because they may develop into serious failures, should be reported at an early stage.

6.7 Maintenance and Testing

Although firefighters are not normally responsible for the routine maintenance of pumps, in the interest of a long trouble-free pump life, attention should be paid to the following:

(i) Strainers must be used where appropriate and checked and cleaned afterwards.

(ii) Following pumping from salt or polluted water the pump, primer, hose reels, water tank and pipework should be thoroughly flushed with clean water to prevent corrosion.

(iii) In cold weather, the pump should be drained of all water after use to prevent possible freezing. If drain plugs need to be removed, they must be replaced securely. Drain cocks should be closed after use.

(iv) On portable pumps if, for any reason, the mixture of water and anti-freeze fluid for the cooling system is spilt onto the pump, or its frame, it should be carefully washed off and dried.

The only requirement for the testing of pumps, at the present time, is that laid down in Technical Bulletin 1/1994. It is a quarterly test for both vehicle mounted and portable pumps and is described as follows:

Quarterly Dry Vacuum Test

> "It is anticipated that this test will normally be conducted by station personnel.
>
> An initial visual inspection of all the lengths of suction hose to undergo testing shall be conducted.
>
> All the lengths of suction hose that have been visually inspected shall then be coupled to the suction inlet of the pump with a blank cap in place on the end of the final length and with the blank caps removed from all deliveries.
>
> The pump shall then be operated at the specified priming speed in order to obtain

0.8 bar vacuum which shall be achieved within 45 seconds.

Having achieved this, the compound gauge shall then not fall to 0.3 bar within 1 minute.

Failure to achieve the above performance indicates an excessive air leak within the system which may be due to leakage at pressure gauge connections, delivery valves, couplings etc."

Any other testing, including pump output tests, may be carried out in accordance with the manufacturer's instructions.

6.8 Assisted Pump and Automatic Tank Fill Controls

6.8.1 Assisted Control Systems (ACS) for Pumps

Although more sophisticated systems are in use abroad, a typical ACS system, as currently used by a number of brigades in the U.K., is simply a means of maintaining a pre-determined pump pressure, under varying demands for water, in order to provide a safe and steady supply to branch operators.

The Home Office Fire Research and Development Group, after evaluation of a number of commercially available overseas systems and discussions with experienced firefighters, produced a draft specification for a system to meet the needs of the U.K. Fire Service. Trials of such a system were conducted in four brigades and the conclusions, which were broadly favourable provided the system could be made reliable, were published, together with a revised specification, in Research Report No.42 (1991).

A number of systems are now commercially available, and, although there are some differences between them, generally speaking they give the following advantages:

(i) **changes in throttle setting, which are normally required to maintain a constant operating pressure when there are variations in water demand, no longer require the intervention of the pump operator. Thus, the problem of pressure surge, which may result from the shutting down of a large jet, and which might endanger other branch operators, is substantially reduced.**

(ii) **delivery pressure regulation is automatic when the supply pressure from a hydrant changes, when changing over from a tank supply to a hydrant supply and when the water level of an open source changes.**

(iii) **a warning can be given if predetermined pump pressure cannot be achieved.**

(iv) **the need for priming is detected and the appropriate pump speed set automatically.**

(v) **the pump operator, whilst still required, is able to perform additional tasks.**

With ACS systems it should always be possible to revert rapidly to manual control if the need arises.

6.8.2 Automatic Tank Fill Controls

An automatic tank fill control allows water from a positive pressure supply, such as a hydrant, to be fed direct to the tank via a valve (see Figure 6.4) which automatically responds to a signal received from level sensors (Figure 6.5). Thus, provided the water demand from the pump (which is normally working from the tank) is not greater than the capacity of the supply, the level within the tank is maintained between specified limits.

An important design feature of such systems, which are more common in European brigades than in Britain, is that the pipework between the tank and the pump should be "full flow" i.e. of large enough diameter to deliver the full rated output of the pump.

The advantages of automatic tank fill controls are:

(i) **variations in supply pressure do not affect branch operators.**

(ii) **the problem of air, contained in long lengths of dry hose, entering the pump**

Figure 6.4 An automatic tank fill control valve.
(Photograph courtesy of West Midlands Fire Service)

and causing significant interruption in supply to branch operators, is eliminated.

(iii) **problems of high inlet pressures which affect some additive induction systems are avoided.**

(iv) **there is always a reserve supply of water in the tank.**

(v) **when it is necessary to reduce branch pressure in an emergency, this can be done simply by use of the throttle. If the pump is fed direct from a hydrant it may be necessary to shut down the delivery as well.**

The disadvantages are:

(i) **it is not possible to take advantage of high hydrant pressure which would normally reduce the throttle setting for a given demand.**

(ii) **there is no obvious indication, except for the tank level gauge, if the demand exceeds the capacity of the incoming supply.**

(iii) **if the appliance is not on level ground, the level sensor may be deceived regarding the actual amount of water in the tank.**

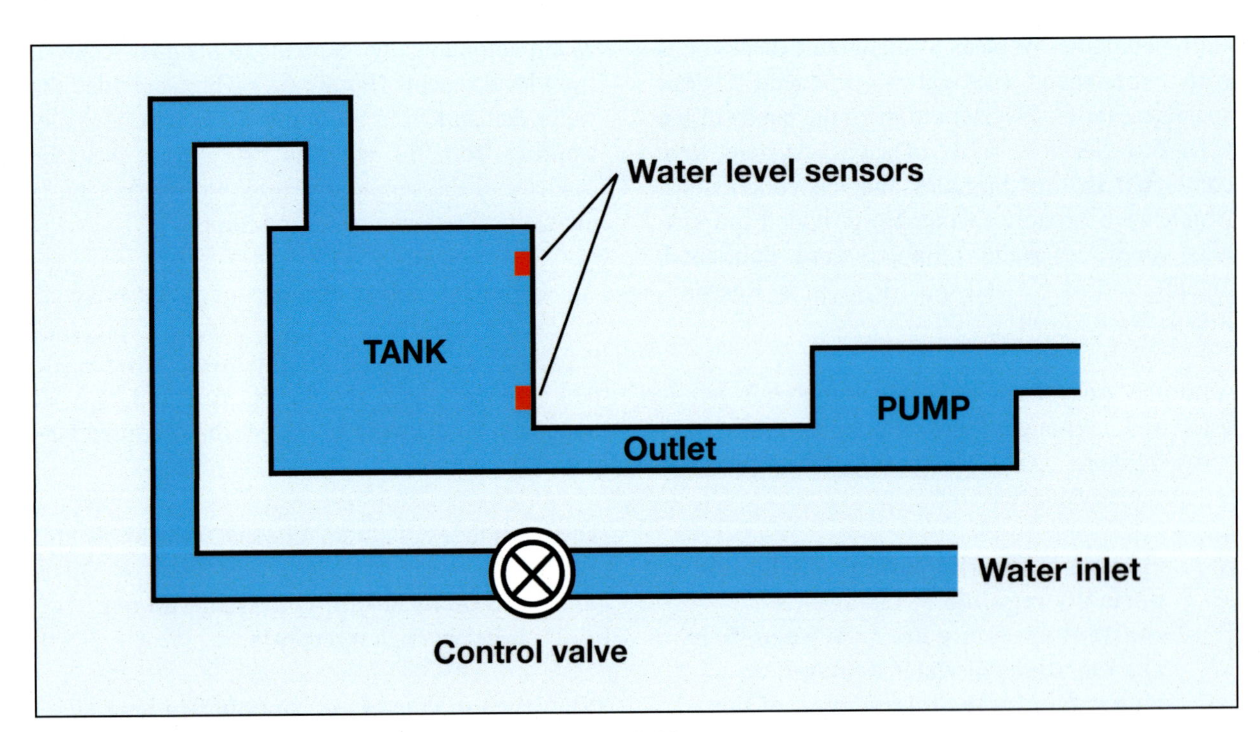

Figure 6.5 Diagrammatic representation of an automatic tank fill system.

Chapter 7 – Pre-Planning

Introduction

The pre-planning of water supplies forms an important part of strategic risk assessment. In many areas mains water supplies are inadequate for firefighting, difficult to access or, in some cases, non-existent. Even where the supply is normally adequate, it may be insufficient to cope with a major incident, e.g. a large industrial fire. The task of providing sufficient water as quickly as possible in such cases needs pre-planning and, sometimes, special equipment. Obviously the first step is to estimate what the water supply needs are likely to be for a particular risk and then, if these requirements exceed the amount immediately available on appliances and from local hydrants, to decide how to make up the deficit. Pre-planning for major fires should involve fire officers at every level of incident command so that an overall strategy can be achieved.

Supplying water to combat an incident can present widely differing problems to different brigades. What constitutes a difficult task to a predominantly rural brigade may not appear to be of such magnitude to a brigade operating in a heavily urbanised area and able to mobilise adequate reinforcing appliances quickly. Nevertheless, the principles of pre-planning still apply. The Central Fire Brigades Advisory Council classifies a 'major' fire as one in which 20 or more jets are required and a survey of such fires has indicated that they generally require at least 17 000 litres/min of water to contain and extinguish them.

7.1 Estimation of Water Requirements

The flowrate required to deal with a particular risk and the period of time for which that flowrate must be sustained depend on many factors. Some of these are listed below.

(i) The extent to which the fire is likely to have spread before firefighting commences.

(ii) The size of the building/area at risk.

(iii) The fire loading.

(iv) Environmental factors – the possibility that nearby water courses may become contaminated.

(v) The construction of the building – materials, compartmentation etc.

(vi) The need for the protection of adjacent risks.

(vii) The value of the building and its contents.

(viii) Whether there is a life risk.

Because there are so many factors to consider, for a large fire risk, it will be difficult to estimate accurately the quantity of water likely to be required for a worst case scenario, but an attempt to do so should be made. There are a number of approaches.

7.1.1 Estimations based on compartment size

A formula used in the United States, the Iowa Rate of Flow Formula, predicts the rate of flow required to control a fire in the largest single compartment of a building when the area is fully involved. When translated into metric units the formula becomes:

$$\text{litres per minute} = \frac{4}{3} \times \text{volume of compartment in cubic metres}$$

The assumption is made that the water is evenly distributed over the surface of the burning contents.

Example 1

What is the rate of flow required to control a fully developed fire in a compartment 20m long, 12m wide and 3m high?

$$\text{Flowrate required} = \frac{4 \times 20 \times 12 \times 3}{3}$$

$$= 960 \text{ litres per minute.}$$

It is not suggested that the use of this formula should be the sole basis for determining required flowrates, but it may be interesting to compare the result with that obtained using a more subjective method.

7.1.2 Estimation based on the number of jets likely to be used

Such an estimate is likely to be based on an experienced firefighter's judgement of how many and what types of branches will be required to deal with the risk in question. If standard, A-type, branches are employed, the flowrate for each may be simply determined using the nozzle discharge formula (derived in Appendix 4):

$$L = \frac{2}{3}d^2\sqrt{P}$$

However, today, diffuser type hand-controlled branches are much more likely to be used and the water requirements of these will depend on the precise models in use with the brigade, their operating pressure and on how they are adjusted (it is possible to set some branches to deliver a particular flowrate).

Some firefighters make an estimate of the water requirements at an incident which is based on the number of lines of hose of the diameter likely to be employed. Practical tests have indicated that, provided hose diameters have been suitably matched to the branches which they supply, and that no more than 4 or 5 lengths are used in each line, the following are reasonable estimates of the likely flowrates:

45mm diameter hose – 300 l/min
70mm diameter hose – 600 l/min
90mm diameter hose – 1200 l/min

7.1.3 Guidelines on Flow Requirements for Firefighting

The LGA/Water UK National Guidance Document on the "Provision of Water for Firefighting" gives the following flowrates as the minima necessary for firefighting in the particular risk categories where new developments are under consideration and suggests their achievement be a condition of planning consent. If additional capacity main is required for firefighting purposes, for which the developer refuses to pay, then the fire authority will have to meet the cost of the water company. In all cases the figures should be regarded only as a guide and the final decision on requirements will be with the fire authority.

(i) **Housing** – from 8 l/sec (480 l/min) for detached or semi detached of not more than two floors up to 35 l/sec (2100 l/min) for units of more than two floors, from any single hydrant on the development.

(ii) **Transportation** – 25 l/sec (1500 l/min) for lorry/coach parks, multi-storey car parks and service stations from any hydrant on the development or within a vehicular distance of 90 metres from the complex.

(iii) **Industry** (industrial estates) – it is recommended that the water supply infrastructure should provide as follows with the mains network on site being normally at least 150mm nominal diameter:

Up to one hectare 20 l/sec (1200 l/min)

One to two hectares 35 l/sec (2100 l/min)

Two to three hectares 50 l/sec (3000 l/min)

Over three hectares 75 l/sec (4500 l/min)

High risk units may require greater flowrates.

(iv) **Shopping, Offices, Recreation and Tourism** – from 20 l/sec (1200 l/min) to 75 l/sec (4500 l/min) depending on the extent and nature of the development.

(v) **Education, Health and Community Facilities**

Village Halls – minimum of 15 l/sec (900 l/min) through any single hydrant on the development or within a vehicular distance of 100 metres from the complex.

Primary Schools and Single Storey Health Centres – minimum of 20 l/sec (1200 l/min) through any single hydrant on the development or within a vehicular distance of 70 metres from the complex.

Secondary Schools, Colleges, Large Health and Community facilities – minimum of 35 l/sec (2100 l/min) through any single hydrant on the development or within a vehicular distance of 70 metres from the complex

7.2 Assessment of Additional Water Supplies

Once the likely water requirements have been established for a particular risk scenario, the question arises as to whether these may be met from the local mains supply or whether other sources such as storage tanks, swimming pools, lakes, reservoirs, rivers and more distant mains will be needed to make up a deficit. The examples which follow make use of some of the formulae for capacity which are derived in Appendix 3.

Example 2

It is anticipated that 4 jets each of which delivers 500 l/min and a ground monitor which delivers 1200 l/min will be required for a particular risk. The local main is capable of delivering only 1000 litres per minute and it is proposed to make up the deficiency from water stored in a cylindrical tank of diameter 8m and depth 1.5m. For how long may firefighting continue before the stored water runs out?

4 jets at 500 litres per minute require 4×500 = 2000 litres per minute.
the ground monitor requirement = 1200 litres per minute
total requirement = 3200 litres per minute

water available from main = 1000 litres per minute
required flowrate from storage tank = 2200 litres per minute

If D is the diameter and h the depth of the tank, both measured in metres, the capacity in litres is given by:

$$\text{capacity} = 800\,D^2h \qquad \text{(Appendix 3)}$$
$$= 800 \times 8 \times 8 \times 1.5$$
$$= 76\,800 \text{ litres}$$

Time for which firefighting may continue $= \dfrac{76800}{2200}$

= 35 minutes approximately

Because this is a relatively short period of time, depending on the nature of the risk it may be considered wise to plan for additional supplies from other, possibly more distant, sources.

Example 3

It is proposed to take water for firefighting from a swimming pool which is 12m long, 4m wide and has a depth which varies uniformly from 0.8m to 2.0m. If the demand is expected to be 800 litres per minute, how long will the supply last?

Capacity of pool in litres

$$= \text{length} \times \text{breadth} \times \text{average depth} \times 1000$$

$$= 12 \times 4 \times \frac{0.8 + 2.0}{2} \times 1000$$

$$= 12 \times 4 \times 1.4 \times 1000$$

$$= 67\,200 \text{ litres}$$

Time for which supply lasts $= \frac{67\,200}{800}$ minutes

$= 84$ minutes

Example 4

A remote country house has a limited mains supply but this may be supplemented, for firefighting purposes, by water from a nearby lake. The estimated surface area of the lake is 1200 square metres and its average depth is 0.7 metres. What is the available supply in litres?

Capacity in cubic metres $= \frac{2}{3} \times$ surface area $\times$ average depth

$= \frac{2}{3} \times 1200 \times 0.7$

$= 560\text{m}^3$

Capacity in litres $= 560\,000$

7.3 Supplying Water to the Fireground

When water has to be conveyed from distant sources to the fireground the two main methods are:

(i) to utilise a number of water tenders or water carriers to maintain a 'shuttle' from the supply;

(ii) to relay the water over the distance using pumps and hose.

When determining an overall policy for water carrying or relaying, the factors which should be taken into account are:

1. The additional quantity of water needed and the time for which it will be required.
2. The location and size of sources, taking into account the time of year and the distance involved for water carrying or relaying.
3. The equipment and methods available for carrying or relaying water.
4. The time factors involved in assembling the necessary equipment.

Once the full picture has evolved, a study of the various options open to the brigade officer can be made. These are discussed in the following paragraphs.

7.4 Water Carrying

In rural areas where distances between water sources and the fireground can be considerable, and where the requirement is for a relatively limited, but continuous, supply of water over a lengthy period, then a system of water carrying may be preferred to a long water relay.

Water carrying can be tackled in one of two ways. Firstly, a number of water tenders can be used to collect water from the source and deliver it either into the tank of a fireground appliance or into a temporary dam. The second method, employed by several brigades, is to use one or more bulk water carriers. A recently conducted survey has shown that these appliances may carry from 3800 litres (a demountable pod) to as much as 13 500 litres, together with a portable or integral pump and a portable dam. The process of delivering water to the fireground and refilling from a source is continued as often as is necessary.

It should be borne in mind that the nearest source of water may not necessarily be the one which should be used, since abundance and ease of access are important factors. This is particularly important when considering the use of water carriers, as a slightly longer transit time may be more than compensated for by a reduction of the time required to complete each filling operation.

The advantages of using water carriers as against a larger number of conventional water tenders are as follows:

1. The total number of appliances required is less.
2. The human resources required are much less.

3. The total time taken to mobilise the appliances can usually be reduced.
4. The number of water-carrying journeys is reduced.
5. Firefighting appliances are not committed merely for carrying water.
6. The operating costs are reduced.

Against these advantages has to be set the capital cost of purchasing and maintaining these special appliances, though leasing arrangements with other owners may be negotiated.

Comparison between the two methods described can be made for an incident on a motorway. A major problem facing an officer in charge of this type of incident is likely to be the 'round trip' distance which appliances need to travel to refill their tanks and return. Because of the distance to access points, a travelling time of 30 minutes is not uncommon in these instances, and so the advantages of water carriers become more apparent.

It is important to appreciate that for water carrying to be practicable, storage arrangements for the delivered water must be available at the fireground. One or more additional water tenders, an inflatable or an improvised dam might be sufficient.

Once estimates have been made for travel, fill and discharge times it is possible to calculate the minimum resources required for a water carrying operation. In practice, one or more additional appliances should be ordered, where possible, to cover any unforeseen delays or other problems.

Example 5

A pre-planning assessment of an incident indicates that a continuous supply of 250 l/min will be required for an indefinite period. The nearest viable hydrant yields 600 l/min and involves a round trip travelling time estimated at 15 minutes. A water tender holds 1800 litres and a water carrier 9000 litres of water. In order to meet fireground requirements:

(a) How many conventional water tenders would be required and how many trips per hour would they have to make?

(b) How many water carriers would be required and how many trips per hour would they be involved in?

(a) A water tender will be able to maintain the required supply for:

$$\frac{1800}{250} = 7 \text{ minutes (approximately)}$$

Time to refill and discharge a water tender with 1800 litres
= 6 minutes (4 min. for fill + 2 min. for discharge)

Round trip travelling time = 15 minutes

Complete trip time = 21 minutes

Thus, each water tender will be able to make approximately 3 trips per hour and, since each tankful lasts 7 minutes, in order to maintain a continuous supply:

$$\text{the number of tenders required} = \frac{21}{7} = 3$$

which, between them, will have to make a total of approximately 9 trips per hour.

(b) A water carrier will be able to maintain the required supply for:

$$\frac{9000}{250} = 36 \text{ minutes}$$

Time to refill and discharge a water carrier with 9000 litres
= 20 minutes (15 min. for fill + 5 min. for discharge)

Round trip travelling time = 15 minutes

Complete trip time = 35 minutes

Thus, to maintain the required supply:

$$\text{the number of carriers required} = \frac{35}{36} = 1$$

which will have to make approximately 2 trips per hour.

This example clearly illustrates the advantages of using water carriers with regard to effective use of personnel and equipment.

The flowrate required in the example (250 l/min) is extremely limited and it is clear that the resources which would be required for more substantial risks will prohibit water carrying, by either water tenders or bulk carriers, as an acceptable option.

7.5 Water Relaying

A water relay comprises a number of pumps spaced at intervals along a route between a water source and the point where the water is required. At one time two types of relay were commonly in use, namely closed circuit (in which the water is pumped through hose direct from one pump to the next) and open circuit (in which it is pumped through hose via portable dams placed between the pumps).

The principal advantage of the open system, i.e. the ability to maintain flow at the fire even if the base pump becomes inoperative, is offset by the greater amount of equipment and effort required. Because of this, the open circuit method is now seldom used and the closed circuit system, which will now be described, is usually adopted. **However, the use of a portable dam at the fireground is recommended because it can act as an emergency supply in the event of relay pump breakdown and also enables better control over jet pressures.**

In a closed circuit relay (Figure 7.1), the first, or base, pump takes its water from the source and pumps through hose lines either directly, or via a series of booster pumps, to the fireground pump. The function of these booster pumps, when they are used, is to compensate for the pressure lost due to friction in the hose. The distance between the pumps is regulated by the amount of friction loss and the contours of the route.

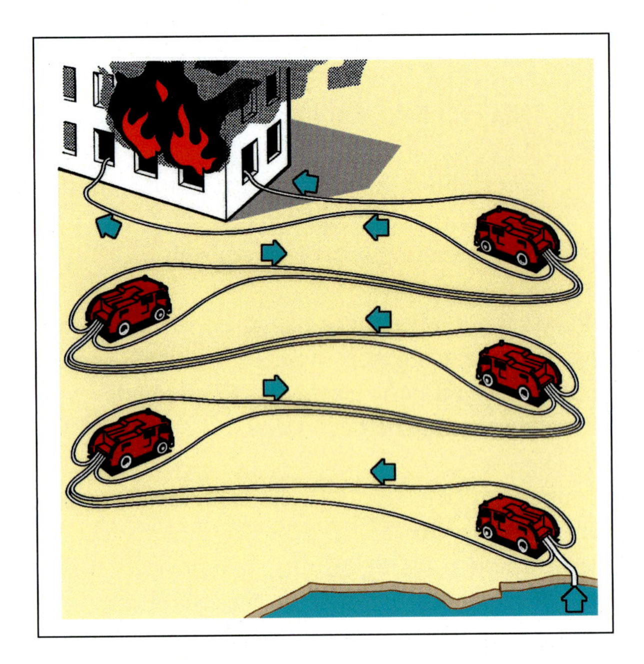

Figure 7.1 Diagram illustrating the closed circuit relay method.

The aim when organising a water relay is to deliver the required quantity of water with the minimum of equipment. Without careful pre-planning it is likely that a greater number of appliances will be used than is strictly necessary, or that the quantity of water delivered will not be sufficient to meet firefighting needs.

The flowrate available from a water relay depends on two factors:

(i) the performance of the pumps used.

(ii) the ability of the hose to convey the water.

A pump's performance can be determined from an inspection of its characteristic or from a knowledge of the maximum flowrate available at a quoted pressure e.g. 7 bar or 10 bar. These issues are discussed fully in Chapter 6. The ability of hose to convey water is limited by friction loss and is therefore highly dependent on its diameter.

The planning of an effective water relay depends on matching the pumps' ability to deliver water with the ability of the hose to convey it. There is no point in having large capacity pumps attempting to deliver water through long lengths of small diameter hose.

For a given flowrate, the maximum spacing between appliances will be determined by the distance over which the pressure in the hose falls to slightly above atmospheric so that it becomes necessary to boost the pressure with another pump. Rather than perform lengthy friction loss calculations to determine this distance, use may be made of the information in Table 7.1a and Table 7.1b.

The information contained in these tables has been obtained from a variety of sources and should be regarded only as a reasonable guide to performance.

Table 7.1a ***Maximum distances for required flowrates between pumps operating at 7 bar for 70mm and 90mm hose with standard instantaneous couplings.***

REQUIRED	MAXIMUM DISTANCE BETWEEN PUMPS (metres) OPERATING AT 7 bar			
FLOWRATE	70mm single	70mm twinned	90mm single	90mm twinned
litres/min	all with standard instantaneous couplings			
400	1500	6000	3900	15700
500	1000	4000	2500	10000
600	690	2700	1700	6900
700	500	2000	1250	5000
800	390	1550	980	3900
900	310	1200	770	3000
1000	250	1000	620	2500
1100	210	820	500	2000
1200	175	690	430	1700
1300	150	590	350	1400
1400	125	500	320	1250
1500	110	440	270	1100
1600	100	390	245	980
1800	75	300	190	775
2000	60	250	155	620
2200	50	205	130	520
2250	50	195	120	490
2500	40	160	100	400
3000	–	110	70	280
3500	–	80	50	200
4000	–	60	40	155
4500	–	–	30	125

Table 7.1b ***Maximum distances for required flowrates between pumps operating at 7 bar for 90mm, 125mm and 150mm hose with Storz couplings.***

REQUIRED FLOWRATE litres/min	MAXIMUM DISTANCE BETWEEN PUMPS (metres) OPERATING AT 7 bar			
	90mm single	90mm twinned	125mm single	150mm single
	all with Storz couplings			
500	3000	12000	17500	50000
1000	750	3000	4700	14000
1500	350	1400	2100	6500
2000	190	780	1200	4000
2250	150	600	950	3100
2500	125	500	750	2500
3000	80	350	550	1800
3500	60	250	400	1350
4000	50	200	300	1000
4500	35	150	250	850
5000	30	125	200	700

Distances may be increased by approximately 40% if pumps operate at 10 bar.

If, for a given flowrate, a line of hose is twinned the flowrate in each line will be halved and consequently the friction loss, which depends on the square of the flowrate, will be reduced to a quarter of its former value. Thus, with twinned lines of hose, the distance between pumps may be four times as great as it is with a single line.

Example 6

Pumps capable of delivering 2500 litres per minute at 7 bar pressure are to be used in a water relay. How many intermediate pumps will be required for this amount of water to be delivered over a distance of 600m using (i) twinned 70mm hose (ii) single 90mm hose with standard couplings (iii) twinned 90mm hose with standard couplings and (iv) single 125mm hose?

Table 7.1 indicates that the maximum distances between pumps and the consequent numbers of intermediate pumps required are as follows:

(i)	for 70mm twinned hose	160m	needing 3 pumps
(ii)	for 90mm single hose	100m	needing 5 pumps
(iii)	for 90mm twinned hose	400m	needing 1 pump
(iv)	for 125mm single hose	750m	none required

Fire service pumps are capable of a variable output according to the discharge pressure at which they are operated. For many pumps currently in use the nominal output is quoted at a pressure of 7 bar and this has been regarded as the normal operating pressure for a water relay. However, it is now becoming common practice for manufacturers to standardise on a pressure of 10 bar when quoting pump performance, and recent improvements in hose construction enable it to cope with this higher pressure, but if a higher standard operating pressure than 7 bar is adopted for a water relay, this will have the effect of increasing the distance possible between pumps at the expense of decreasing the amount of water available. What is regarded as the optimum arrangement for a relay occurs when the distance between pumps is the maximum which will allow them to deliver their full rated output through the hose selected and this may easily be determined from Table 7.1 as has been shown in the example above. However, the water demands at an incident may not be such as to require the full rated output of the pumps, in which case it becomes possible to operate with longer stages.

Example 7

If a water relay is required to deliver 1000 litres per minute at an incident, assuming a pumping pressure of 7 bar, what is the maximum distance between stages using (i) twinned 70mm hose (ii) twinned 90mm hose with Storz couplings and (iii) single 125mm hose?

Table 7.1 indicates that the distances are as follows:

(i) for 70mm twinned hose – 1000m

(ii) for 90mm twinned hose – 3000m

(iii) for 125mm single hose – 4700m

The only requirement regarding the pumps is that they should be able to maintain a flowrate of at least 1000 l/min when operating at 7 bar pressure.

Examples 6 and 7 illustrate clearly the advantages, from the resource point of view, of twinning lines and increasing hose diameter. However, if the water supply is of limited capacity, some thought should be given as to just how much of the scarce resource is required simply to fill the hose.

Example 8

What is the volume of water contained in 1000m of 125mm hose?

The capacity of 1m of hose is given by the formula:

$$\text{capacity} = \frac{8\ d^2}{10\ 000} \qquad \text{(appendix 3)}$$

so the required capacity is:

$$\frac{8 \times 125 \times 125 \times 1000}{10\ 000}$$

$$= 12\ 500 \text{ litres}$$

Ideally the pumps used in a relay should all have the same capacity and, with the possible exception of the first two, should be equally spaced.

If pumps of different capacities are used, the maximum output of the relay will be dictated by the output of the pump with the lowest capacity.

If the pumps are not equally spaced, the maximum output of the relay will be dictated by the flowrate in the longest stage.

7.6 Practical Considerations

7.6.1 Relaying over undulating ground

It frequently happens that the ground over which a relay is laid is not level, and if the gradients concerned are sufficiently great some adjustment of the distances between pumps becomes necessary.

Where the ground between pumps is uphill, some of the pump pressure will be used up in raising the water to the higher level with the result that less pressure will be available to overcome frictional resistance in the hose. It may be possible to increase the pump pressure in order to compensate for this, provided that the pump is not already working at full throttle, but otherwise the intended flowrate can only be maintained by reducing the distance between the pumps. Conversely, where the ground between pumps lies downhill, the static head offsets the frictional resistance in the hose, so the pumps can be spaced farther apart.

An appropriate method for determining the spacing is to estimate the difference in height between adjacent pumps in metres and divide this by 10 to give the static head requirement in bar. This figure should then be divided by the operating pressure of the relay and the distance between pumps reduced or increased by the resulting fraction.

Example 9

A section of a relay, for which the appropriate spacing for pumps operating at 7 bar on level ground is 200m, has to be laid up a hill with a gradient such that the difference in level over this distance is estimated to be 20m. What should be the spacing on the hill?

The pressure equivalent to 20m head of water is 2 bar, so the effective pressure for overcoming friction is reduced to:

$$7 - 2 = 5 \text{ bar}$$

In order to maintain the flowrate, the spacing should therefore be reduced by:

$$\frac{2}{7} \times 200 = 57\text{m}$$

i.e. to 143m

It should be noted, however, that where the difference in level between successive pumps is less than 10 metres, the percentage variation in flowrate will be in single figures, so differences in level of less than 10 metres may normally be disregarded.

If the relay has to be laid downhill on a similar gradient, the distance between pumps could be increased by:

$$\frac{2}{7} \times 200 = 57\text{m}$$

i.e. to 257m

7.6.2 Position of the base pump

The output from the base pump, which has to supply the water, will control the flow through the relay. If this pump is not working efficiently, the whole of the relay will be impaired.

When working from open water, suction conditions will govern the input of the base pump; the full rated output can be expected if the vertical suction lift is not more than 3 metres and no more than three lengths of suction hose are used. The base pump should therefore be situated as near as practicable to the water source, with the minimum suction lift.

7.6.3 Spacing between first two pumps

Because a base pump working from open water has to use a part of its energy in lifting the water from the source to the pump inlet, there will be something of a reduction in the pressure available to pump the water through the hose on the delivery side to the first booster pump. In such circumstances it may be appropriate to reduce the distance between the base pump and the first booster pump to compensate for this loss of pressure.

7.6.4 Communications

For the efficient operation of a water relay, it is important to maintain good communications along the route, so that changes in conditions, orders to shut down, etc. can be acted upon quickly. The type of communications adopted will, of course, be dependent on conditions, availability of resources, and so on, and it is the responsibility of the relay officer to devise an appropriate system. If using radio sets, particularly at large incidents, one channel should be dedicated solely for use by water relay officers and operators. Whatever system is used, the communications operators should be positioned a short distance away from the pumps because of the noise of the engines.

7.6.5 Charging with water

Provided good communications have been established, it may be advantageous, especially if large diameter hose is being used, to partially charge the hose as the relay is being assembled. One delivery on each booster pump should be left open, in addition to the deliveries connected to the hose lines, to facilitate the removal of air from the system. As soon as water reaches the pump, this extra delivery should be closed.

The base pump should run at about half speed until the whole system has been charged. When the officer in charge is satisfied that the relay is working satisfactorily, the speed of the base pump should be gradually increased until the full pressure is reached. During this period and subsequently, the booster pump operators should keep the relay in balance by gradually adjusting their throttles to a position where the compound gauge is reading just above zero. Booster pumps, having no suction conditions to contend with, should be running at a slightly slower speed than the base pump.

7.6.6 The Porter Relay

This incorporates a procedure to quickly establish a relay when the number of appliances initially available is limited and, as more appliances arrive, to gradually increase its capacity, but without at any time needing to interrupt the water supply. The arrangement was originally conceived for 2250 l/min pumps each of which carried 14 lengths of 70mm hose. However there is no reason why the principle could not be adapted to work with other equipment. Figure 7.2 shows the three stages in the building up of the relay to its full capacity, starting with just 3 pumps and a single line of hose but ending up with 5 pumps and twinned hose.

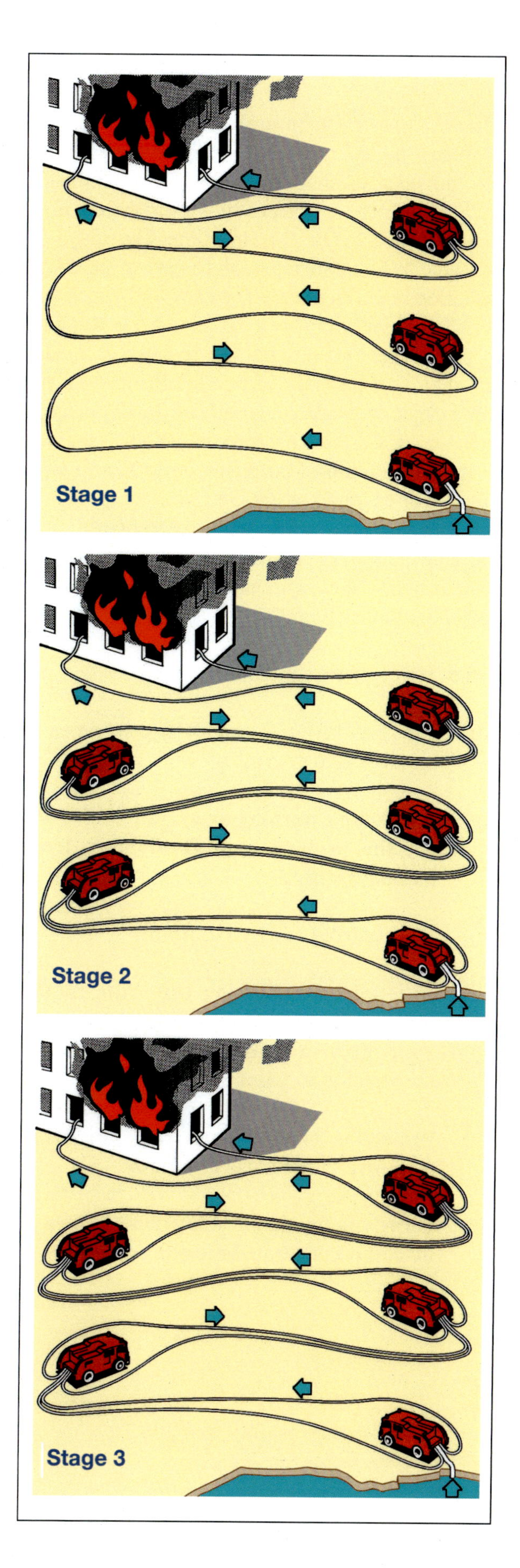

Figure 7.2 The stages of the Porter Relay.
Stage 1: *14 lengths of single 70mm hose between appliances.*
Stage 2: *Intermediate appliances positioned when they arrive and second line deployed without disturbing first line.*
Stage 3: *7 lengths of twinned 70mm hose between all appliances.*

7.6.7 Mechanical Breakdowns

Should a booster pump suffer a mechanical defect, provided the lines have been twinned there is usually no need to shut down the relay completely because when a replacement pump arrives the lines may be connected to it one at a time. The relay will continue to function, although of course there will be a drop in the output and the throttles of the other booster pumps will have to be adjusted in the light of the changed conditions. The incident commander should, if possible, have a spare pump of the same capacity available, with crew, ready to set in to the relay. Careful consideration should be given to the substitution of portable pumps for a major pump. For example, if a 2250 l/min pump operating in a twin line relay has to be replaced with portable pumps then, if the performance of the relay is not to be compromised, one will need to be substituted in each line.

7.6.8 Safety Precautions

Among the factors which should be considered in the interests of safe systems of work are:

(i) **Positioning warning signs and coning off for the protection of firefighters from passing traffic.**

(ii) **The wearing of hi-viz clothing.**

(iii) **Carefully pre-planning the route of the relay and positioning the hose lines and appliances so as to cause the minimum obstruction to passing traffic.**

(iv) **Protection of personnel from long-term exposure to the noise of pumps.**

(v) **Clearly defined and well practised means of communication between relay operators.**

7.7 Special Equipment

7.7.1 Use of Hose Layers

When hose is being laid direct from a hose-laying lorry, the vehicle should be driven at a steady 15 to 25km/h. This speed may be increased to a maximum of about 40km/h in ideal conditions, but extreme care must be taken as, at this speed, the hose is likely to over-run when the vehicle slows down, particularly at corners. This is likely to cause unnecessarily large bights of hose which would result in excessive snaking and kinks when the hose is charged.

If possible, after the hose has been laid and before it is charged, a check should be made to see that it is lying at the side of the road and is causing no obstruction to traffic. The opportunity should be taken to remove any kinks before charging.

When it is necessary to take hose across a road, ramps or bridging units should be used.

Some Brigades have water tenders fitted with hose 'coffins' which store flaked hose within a side locker at the rear of the fire appliance. Hose is deployed in the same way as with a specialist hose layer, enabling the first line of a water relay to be quickly established. The hose 'coffin' is mounted on rollers and can be easily pulled free from the locker, as shown in Figure 7.3, thus enabling the hose to be restowed after use.

7.7.2 Deployment and Retrieval of Large Diameter (Hi-Vol) Hose

Hose layers for the rapid deployment of standard lengths of 70mm and 90mm hose have been in use for some time and, because each length is relatively flexible and lightweight, retrieval of the hose after an incident presents no great problems. The distinct advantage of larger diameter hose for water relaying has been made clear earlier in this chapter but, because its weight increases in proportion to its diameter and, because, for speed of deployment and hydraulic efficiency, it is made in much longer lengths, the problem of retrieval is much greater. It is for this reason that systems have recently been developed for the rapid deployment and retrieval of this larger diameter hose.

Figure 7.3 Storage of hose for rapid deployment.
(Courtesy of West Midlands Fire Service)

Careful consideration should be given at, the pre-planning stage, to the route over which large diameter hose is to be laid, particularly where it crosses roads and in the area around the fireground, because once charged, it presents an obstruction to the movement of fire appliances and other vehicles. Suitable large hose ramps are available but an alternative solution to the problem of obstruction of roads etc. is to use, in the problem areas, a number of lines of 90mm hose with appropriate dividing and collecting breechings. However, this procedure will to some extent reduce the hydraulic efficiency of the relay.

A hose deployment and retrieval unit currently used in brigades and capable, with manual assistance, of retrieving up to 1000m of 125mm or 150mm flaked hose directly back onto the hose laying vehicle ready for re-use is shown in Figure 7.4.

Figure 7.5 shows a hose laying and retrieval system, recently introduced by West Sussex Fire Brigade, which is based on systems extensively used in the United States. It consists of two large hydraulically driven hose drums capable of carrying a total of 1200 metres of 150mm diameter hose

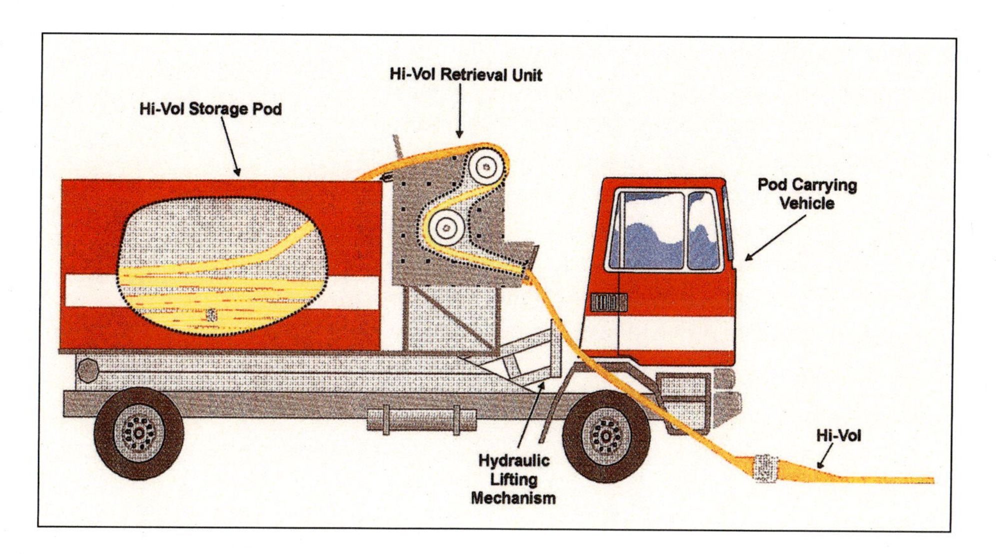

Figure 7.4 A flaked hose deployment and retrieval unit.
(Diagram courtesy of Angus Fire)

Figure 7.5 The West Sussex Fire Brigade hose laying unit.
(Photo: West Sussex Fire Brigade)

and has the advantage that all deployment and retrieval operations may be carried out with personnel at ground level. Closed circuit television enables the driver of the unit to monitor both the operation of the drums during deployment and retrieval and also, in the interest of safety, the movement of personnel around the rear of the vehicle. A forklift truck, carried at the rear of the unit, is provided to assist with the movement of the drums and for the handling of pallets of additional flaked hose and hose fittings.

Before commencing retrieval of large diameter hose, water should be allowed to free flow out of the hose and, where water sits in hollows on undulating ground, the couplings should be broken, the hose drained and the couplings re-attached. Because Hi-vol hose is manufactured in long lengths, the pressure due to the head of water at a coupling near the bottom of a hill may be such as to make it difficult to break. The insertion of a dividing/collecting breeching (Figure 7.10) at such points will facilitate drainage from the hose and so allow the coupling to be broken and the device removed before retrieval.

The driver of the retrieval unit should move forward at a speed similar to the rate at which the hose is being collected in order to avoid excess tension on, or over-running of, the hose line.

If personnel are required to access the flatbed of a hose laying unit during deployment and retrieval operations then, in order to minimise the risk of accidents, an appropriate safe system of work must be devised.

7.7.3 The Holland Fire System for Pumping from Open Water

It has already been explained in Chapter 3 that the performance of a typical pump is significantly reduced when operating at a lift of more than a few metres. One way of overcoming this problem is to employ a hydraulically driven pump which is immersed in, or floats on, the water supply. The Holland system uses 150mm diameter hose and depending on the lift required (which may be up to 60m) and the distance to the fireground, is able to pump up to 4250 litres per minute. It is a containerised system consisting of:

(i) a hydraulically driven portable submersible pump.

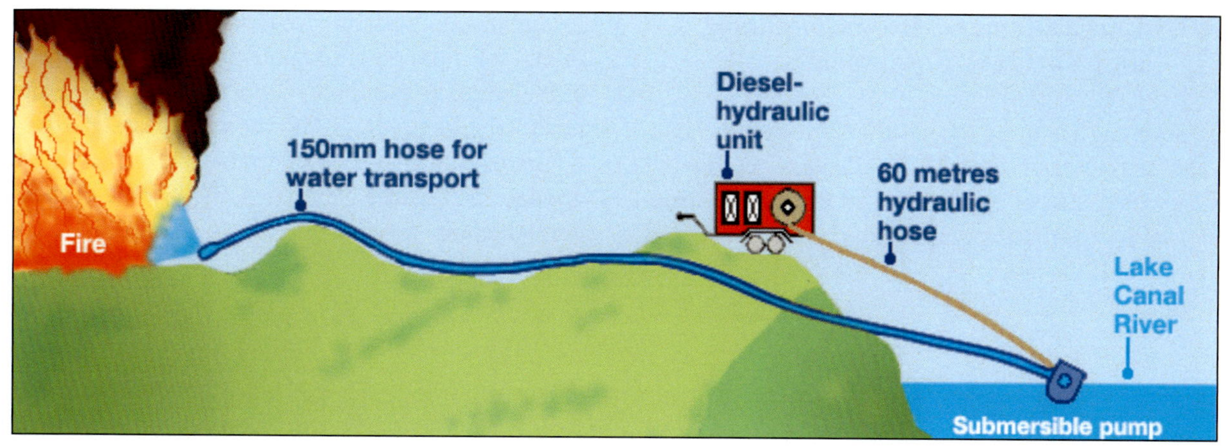

Figure 7.6 The Holland Fire System pumping from an open source.

(ii) a 60m long transmission system consisting of hydraulic hose capable of operating at pressures up to 320 bar.

(iii) a 133kW diesel powered hydraulic drive unit.

(iv) a hosebox containing 1km of 150mm hose (units capable of storing up to 3km are available).

(v) a truck for the transportation of all of the above equipment.

Figure 7.6 shows, diagrammatically, how the equipment is deployed.

Figure 7.7 shows the submersible pump, drive unit, and the truck mounted hosebox and hose retrieval system.

Figure 7.7 The Components of the Holland Fire System.
(Photo: Kuiken Hytrans)

7.7.4 Large Diameter Hose Couplings and Ancillary Equipment

If advantage is to be taken of large diameter hose, then appropriate couplings, adapters, collecting heads etc. must be used. Since, at the present time, pumping appliances are fitted with standard instantaneous coupling outlets and screw thread inlets for hard suction, adapters are needed in both cases to facilitate connection to the Storz hose couplings. It is therefore necessary for a component box of such equipment to be stored on appliances. Figures 7.8 to 7.11 show some of these components.

Figure 7.8 An intermediate pump component box with, on the left, Storz to screw thread adapters for connection to the pump inlet.
(Photo: Northamptonshire Fire and Rescue Service)

Figure 7.9 A Storz to instantaneous male collecting breeching to facilitate connection to the pump outlets via a number of short lengths of 70 or 90mm hose.
(Photo: Angus Fire)

Figure 7.10 A dividing/collecting breeching for use when multiple lines are required. *(Photo: Angus Fire)*

Figure 7.11 A 'Phantom Pumper' which enables hose with standard couplings to be connected directly to the large diameter line.
(Photo: Angus Fire)

7.7.5 Use of Helicopters

Though used only by a limited number of brigades, in particular those with a responsibility for large areas of forest or moorland, helicopters have in the past few years proved to be extremely effective in situations where access to the fire is difficult for both firefighters and firefighting equipment. In addition to their use for aerial observation and the transportation of personnel and equipment, they may also be used to transport water to a point close to the seat of the fire or to discharge water directly onto the fire.

For the transportation of water to an incident, a large bucket suspended beneath the helicopter and capable of containing several hundred litres of water is used. This can rapidly be filled from any sufficiently large open source, including an appropriately designed portable dam with a minimum depth of 1.2 metres, and, by opening a remotely controlled electrical valve situated at the base of the bucket, water may be discharged either directly onto the fire or into a portable dam (Figure 7.13) or all-terrain vehicle. The use of helicopters for the laying of hose from an underslung drum, although an attractive proposition, has been ruled out by the civil aviation authority because of the danger of the hose snagging and the risk of crossing high voltage power lines.

A helicopter is an extremely expensive item of equipment to purchase outright and even leasing costs may amount to several hundred pounds per hour. However, its ability to provide rapid intervention at incidents such as forest fires, where

Figure 7.12 A helicopter collecting from open water using an underslung bucket.

Figure 7.13 Another discharging into a portable dam.

(Courtesy Derbyshire Fire and Rescue Service)

access for wheeled vehicles would be difficult and time consuming, may well result in the early conclusion of firefighting procedures which might otherwise prove to be very prolonged with proportionally high costs in manpower and equipment.

At the time of writing, further operational trials to establish the full potential of helicopters for fire service applications are being conducted.

Hydraulics, Pumps and Water Supplies

Glossary of Hydraulics Terms

Base pump — The first pump in a water relay, taking its water direct from the source.

Booster pump

1. A pump used to increase the pressure in a water main.
2. In a water relay, any pump set in between the base pump and fireground pump.

Casing — The chamber surrounding the impeller in a centrifugal pump.

Cavitation — The formation of bubbles of water vapour in a pump, caused by the water boiling at low pressure.

Centrifugal pump — A pump in which water is moved by the spinning action of an impeller.

Closed circuit

1. Cooling system: one in which all the cooling water from the fire pump is returned to the pump, none of it being discharged to waste.
2. Water relay: one in which the water is pumped direct from one pump to another.

Collecting breeching — A connection designed to join two lines of hose into one.

Compound gauge — A gauge designed to measure both positive and negative pressures.

Delivery — On a pump, the valved outlet through which water is discharged.

Diffuser — A system of guide vanes in the casing of a pump to reduce turbulence.

Ejector pump — A pump which lifts water by means of a partial vacuum created by a jet of water under pressure from another pump.

Friction factor — A figure expressing the degree of internal roughness of a particular type of hose or pipe. It must be taken into account when calculating friction loss.

Friction loss	Loss of water pressure caused by the frictional effect of the walls of the hose or pipe through which the water passes.
Full-flow coupling	A type of coupling which allows water to pass from one length of hose to another without obstruction or reduction of diameter.
Gland	A device fitted around a shaft, eg in a pump or hydrant, which exerts pressure in order to prevent leakage.
Governor	A device which automatically controls the speed of a pump engine in order to maintain a preset pump pressure.
Guide ring	A ring of guide vanes in the casing of a centrifugal pump designed to reduce turbulence.
Hard suction	Large-diameter hose internally braced to prevent collapse under atmospheric pressure.
Head	1. The vertical distance from a given point in a fluid to the open surface of that fluid. 2. The depth of open water equivalent to a given pressure.
Heat exchanger	A cooling device in which heat is conducted from the hot substance to a cool circulating fluid.
Hi Vol	The name given to hose with a diameter in excess of 90mm.
Hydrostatic	To do with forces arising from a stationary fluid.
Impeller	The spinning part of a centrifugal pump which imparts a high velocity to the water.
Instantaneous	A type of interlocking coupling designed to be connected by pressing onto a seal (as opposed to being screwed in).
JCAEU	Joint Committee on Appliances, Equipment and Uniform (a committee of the Central Fire Brigades Advisory Council). Formerly JCDD.
JCDD	Joint Committee on Design and Development of Appliances and Equipment.
Kinetic	To do with motion.
LWP	Lightweight portable pump.
Mechanical seal	A seal which uses a spring to maintain contact, instead of packing material.

Monitor — A piece of equipment for delivering very large quantities of water. It may be freestanding (ground monitor) or appliance mounted (eg TL monitor).

Multi-pressure pump — A centrifugal pump which combines a conventional low pressure stage with a high-pressure stage.

Multi-stage pump — A centrifugal pump with two or more impellers.

Negative pressure — Pressure lower than that of the atmosphere (atmospheric pressure being regarded as 'zero' in this context).

Non-percolating hose — Any non-porous hose.

Open circuit — **1.** Cooling system: one in which the cooling water from the fire pump discharges to waste. (This system is no longer used on fire appliances in the United Kingdom.)

Open circuit — **2.** Water relay: one in which the water is pumped via portable dams placed between pumps. (Seldom used nowadays.)

Percolating hose — Any unlined, porous hose.

Peripheral pump — A special type of centrifugal pump in which the water follows a spiral path around the edge of the impeller.

Positive pressure — Pressure higher than that of the atmosphere (atmospheric pressure being regarded as 'zero').

Power take-off (PTO) — A device to divert engine power from running an appliance to running equipment on it, such as a built-in pump.

Pressure — **1.** Force exerted per unit of area.

2. In practical firefighting: same as positive pressure.

Pressure gauge — A gauge designed to measure positive pressure only.

Primer — A device for filling a centrifugal pump with water to enable it to operate from a non-pressure-fed supply.

PTO — Power take-off.

Rack valve standpipe — A type of standpipe with a valve at the head enabling the flow to be shut off.

Reciprocating pump — A pump in which water is moved by the action of a piston or plunger in a cylinder.

Seal — Another name for a gland.

Shuttle — A system in which a number of water tenders or water carrier are used in rotation to convey water to the fireground.

Single-stage pump — A centrifugal pump with one impeller.

Sluice valve

1. A valve in a water main, used to shut down the main if necessary, or to divert water.

2. A valve found in one type of hydrant.

Soft suction — Non-reinforced hose designed for use in positive pressure situations.

Static pressure — The pressure at a hydrant or pump when the water in it is stationary.

Static suction lift — The vertical distance from the surface of an open water source to the eye of the pump.

Statutory water undertaker — Any regional water authority or statutory water company.

Stuffing box — A chamber, eg in a hydrant, on which pressure is exerted by a gland in order to prevent leakage around a spindle, shaft etc.

Tachometer — An instrument for measuring the speed of an engine.

Tuberculation — The roughening of the internal surface of a pipe due to the build up of deposits.

Twinning — The laying of two lines of hose along the same path.

Vacuum

1. An empty space, ie one containing no matter.

2. In practical firefighting: same as negative pressure.

Vacuum gauge — A gauge designed to measure negative pressure only.

Volute — A type of casing in a centrifugal pump, shaped like the shell of a snail, where kinetic energy is converted to pressure energy.

Vortex — A swirling depression in the water surface caused by the drawing down of air when a suction strainer is insufficiently submerged.

Water carrier — A vehicle used for conveying large quantities of water to an incident where it is difficult to obtain an adequate supply otherwise.

Appendices

APPENDIX 1

Symbols and Units

Throughout this Publication various symbols are used to denote pressures, dimensions, etc. The symbols are shown below and should not be confused with standard abbreviations for SI units which are given in the Guide to SI units.

A = cross-sectional area (square metres)
a = cross-sectional area (square millimetres)
b = breadth (metres)
BP = brake power (watts)
C = circumference (metres)
D = diameter (metres)
d = diameter (millimetres)
E = efficiency (per cent)
F = force (newtons)
f = friction factor
g = acceleration due to gravity (= 9.81 m/s^2)
H = metres head
h = depth (metres)
L = flow (litres per minute)
l = length (metres)
m = mass (kilograms)
p = pressure (newton/square metre)
P = pressure (bar)
P_f = pressure lost in friction (bar)
Q = flow (cubic metres per second)
R = reaction (newtons)
r = radius
s = sum of the sides of a triangle
t = time (seconds)
v = velocity (metres per second)
WP = water power (watts)
ρ = density (kilograms/cubic metre) (*Greek symbol ro*)
π = 3.1416, or $^{22}/_{7}$ (*Greek symbol pi*)
= is equal to
≃ is approximately equal to [Can also appear as ≏ or ≒]
∴ therefore

APPENDIX 2

Transposition of Formulae

A.2.1 The need for Transposition

In order to illustrate the need for transposition let us take a typical hydraulics formula such as:

$$WP = \frac{100LP}{60}$$

which enables us to calculate the water power (WP) of a pump when we know the discharge (L) from it and the pressure (P) at which it is operating.

The term on its own on the left of the equals sign, in this case WP, is called the subject of the formula.

It is just as likely, however, that we might wish to determine L given the numerical values for P and WP. What we then need, ideally, is a re-arranged version of the formula in which L is the subject (i.e. appears on its own on the left of the equals sign) and with all the other terms on the right. The process of re-arrangement, which results in a new subject of the formula, is called transposition.

In the example chosen, because there are three variables (WP, P and L) there will be three possible ways of arranging the formula, and the choice we have is between memorising all three versions or being able to perform simple transposition on the one normally remembered in order to derive the other two. Bearing in mind the proliferation of hydraulics and other formulae where the problem might occur, the acquisition of the ability to apply the limited number of rules required to transpose the great majority of formulae would appear to be the better alternative.

A.2.2 The basic rules for Transposition

Let us return to the example in the above paragraph where the problem posed was to re-arrange the Water Power formula so as to make L the subject. Our objective may be considered as twofold: (i) to get L on the left of the equals sign and (ii) to get all the other symbols on the right.

The equals sign in any formula is making the statement that, when appropriate numbers are substituted for the symbols and any subsequent arithmetic performed, the two sides will reduce to the same numerical value. It must, therefore, be equally true to make the same statement in reverse

Thus:

$$\frac{100LP}{60} = WP \qquad \textit{stage 1}$$

Though not a particularly spectacular move, we have achieved the first part of our objective – L is now on the left of the equals sign.

Again, if two numbers are equal, the results of multiplying (or dividing) them both by the same quantity will also be equal. We may therefore multiply both sides of the above formula by 60 if we wish.

Let us see the result of doing this.

$$\frac{60 \times 100LP}{60} = 60 \times WP$$

Clearly the '60's will now cancel out on the left-hand side leaving us with:

$$100LP = 60WP \qquad \textit{stage 2}$$

(note that there is no need to retain the line under the left-hand side since there are no symbols remaining in the denominator)

Now let us divide both sides of the formula by 100:

$$\frac{100LP}{100} = \frac{60WP}{100}$$

Again the '100's cancel out on the left leaving:

$$LP = \frac{60WP}{100} \qquad \textit{stage 3}$$

Finally let us divide both sides of the above formula by P:

$$\frac{LP}{P} = \frac{60WP}{100P}$$

Cancelling the 'P's on the left gives:

$$L = \frac{60WP}{100P} \qquad \textit{stage 4}$$

and we have finally achieved our twofold objective with L on its own and on the left of the equals sign.

If we now compare our newly transposed formula (stage 4) with the reversed form of the original formula (stage 1) we can see that the net result of the intermediate stages is that numbers and symbols which we wished to remove from the left-hand side ('100' and 'P' and '60') have moved diagonally across the equals sign; those which were above the line on the left at stage 1 ('100' and 'P') appear below it when transferred to the right at stage 4 and the '60', which was below the line at stage 1, appears above it when transferred to the right at stage 4.

The following two rules summarise the processes described above:

1. **We may reverse the formula.**

2. **We may transfer symbols and numbers from one side of the formula to the other provided we move them diagonally across the equals sign.**

A.2.3 Applying the rules

We will consider a number of different cases.

Example 1

Transpose the formula for the percentage efficiency of a pump:

$$E = \frac{WP}{BP} \times 100$$

to make WP the subject.

In this example the proposed new subject is above the line and uncomplicated by the presence of roots or indices.

Applying rule 1(reversing the formula) gives:

$$\frac{WP}{BP} \times 100 = E$$

Applying rule 2 (moving the unwanted symbols on the left diagonally) gives:

$$WP = \frac{E \times BP}{100}$$

Example 2

Transpose the formula for the velocity of a falling body:

$$v = gt$$

to make t the subject.

In this case there is no line on the right hand side, because there are no symbols in the denominator, but the rules are applied in exactly the same way.

Applying rule 1 gives:

$$gt = v$$

Applying rule 2 gives:

$$t = \frac{v}{g}$$

Example 3

Transpose the formula for force:

$$F = \frac{mv}{t}$$

to make t the subject.

In this example the proposed new subject is *below* the line and we do not apply rule 1

Applying rule 2 to t (moving it diagonally) gives:

$$tF = mv$$

(note that the order of the symbols does not matter but if numbers are involved we put them first)

Applying rule 2 to F gives:

$$t = \frac{mv}{F}$$

Both of these steps may be carried out at the same time of course.

A2.4 Transposition where roots or indices are involved

Example 4

Transpose the formula for nozzle discharge:

$$L = \frac{2d^2 \sqrt{P}}{3}$$

to make P the subject.

The formula is best re-written as:

$$L = \frac{2d^2 \sqrt{P}}{3}$$

to make clear what is above and what is below the line.

The proposed new subject, P, appears in the formula under a square root sign.

However, it is above the line, so the first step is the usual one of reversing the formula:

$$\frac{2d^2\sqrt{P}}{3} = L$$

Now we apply rule 2 to leave only $\sqrt{P}$ on the left. Make no attempt, at this stage, to remove the root sign:

$$\sqrt{P} = \frac{3L}{2d^2}$$

Finally, to remove the root sign, we multiply each side by itself:

$$\sqrt{P} \times \sqrt{P} = \frac{3L}{2d^2} \times \frac{3L}{2d^2}$$

$$P = \left(\frac{3L}{2d^2}\right)^2$$

It is extremely important to include the brackets around the terms on the right. Without them the power 2 would apply only to L.

Example 5

Transpose the formula for friction loss:

$$P_f = \frac{9000fl L^2}{d^5}$$

to make L the subject.

The proposed new subject, L, is raised to the power 2.

However it is above the line, so the first step is the usual one of reversing the formula:

$$\frac{9000fl L^2}{d^5} = P_f$$

Now apply rule 2 to leave only L^2 on the left. Make no attempt, at this stage, to remove the power 2:

$$L^2 = \frac{P_f d^5}{9000fl}$$

Finally, to remove the power 2, we take the square root of each side:

$$L = \sqrt{\frac{P_f d^5}{9000fl}}$$

A.2.5 Summary

Application of the rules described above enables us to transpose the great majority (but not all) of formulae likely to be encountered in the field of hydraulics and other areas.

They may be summarised as follows:

1. **The first step is usually to reverse the formula in order to get the new subject on the left of the equals sign. An exception to this is when the new subject is below the line, in which case go straight to step 2.**

2. **Move numbers and symbols diagonally across the equals sign to leave only the new subject on the left. Do not attempt to deal with any root signs or indices at this stage.**

3. **Squaring both sides will eliminate root signs.**

4. **Taking the square roots of both sides will eliminate powers of 2.**

Calculation of Areas and Volumes

The calculation of areas and volumes, especially of rectangular and circular shapes, is a fundamental part of the mathematics required by firefighters when making calculations involving water resources and requirements. It is proposed, therefore, in this appendix to deal with the general principles of measurement.

A.3.1 Area of regular figures

The area of a rectangle or triangle is easily calculated, as shown in Figure 1.

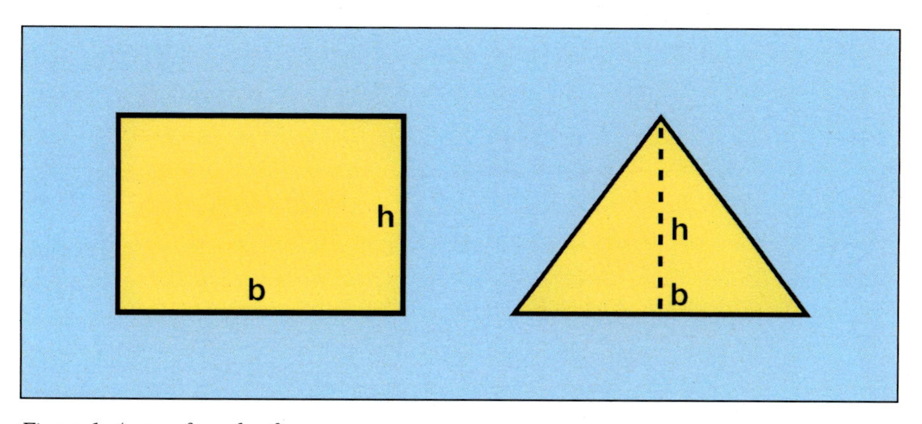

Figure 1 Areas of regular figures

Area of rectangle or square = **base** (b) **× vertical height** (h)

Area of a triangle $= \dfrac{\textbf{base (b) × vertical height (h)}}{2}$

The area of a triangle can also be calculated when the length of each side is known but when it is impossible to measure the vertical height, as for example, in the case of a large triangular-shaped pond or lake. The required formula is:

Area of a triangle $= \sqrt{s\,(s-a)(s-b)(s-c)}$

where **s** = one half of the sum of the sides, a, b and c.

A.3.2 Area of irregular figures

To find the area of irregular figures it is generally necessary to divide them into regular shapes such as rectangles or triangles and find the area of each.

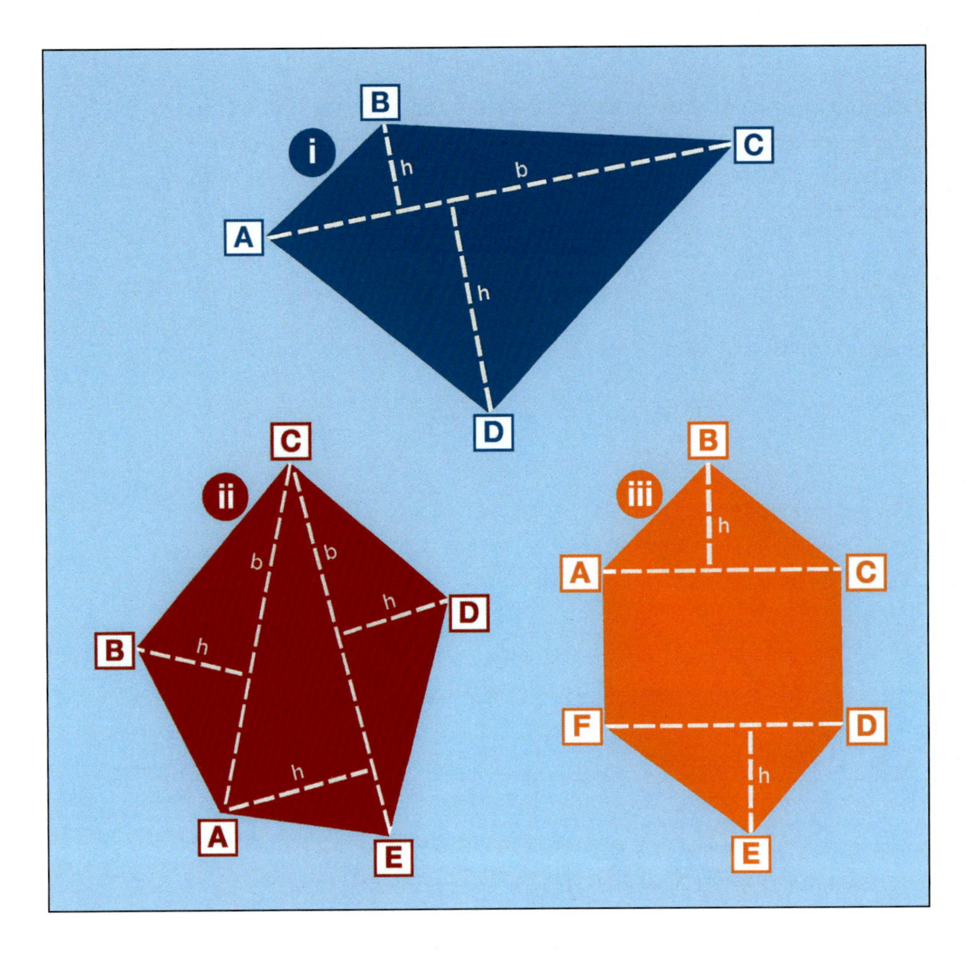

Figure 2 Areas of irregular figures.

The area of the quadrilateral ABCD (Figure 2 i) can be found by dividing it into two triangles ABC and ACD, and adding together the area of each.

The figure ABCDE (Figure 2 ii) can be divided into three triangles, ABC, ACE and CDE, whilst the figure ABCDEF (Figure 2 iii) can be divided into a rectangle ACDF, and two triangles ABC and DEF.

If it is required to ascertain the surface area of an irregularly-shaped pond or lake, this can be calculated by sketching its shape (Figure 3(1)), drawing a rectangle ABCD around it and then inserting such rectangles and triangles as will roughly fill the area outside the water surface, but inside the rectangle.

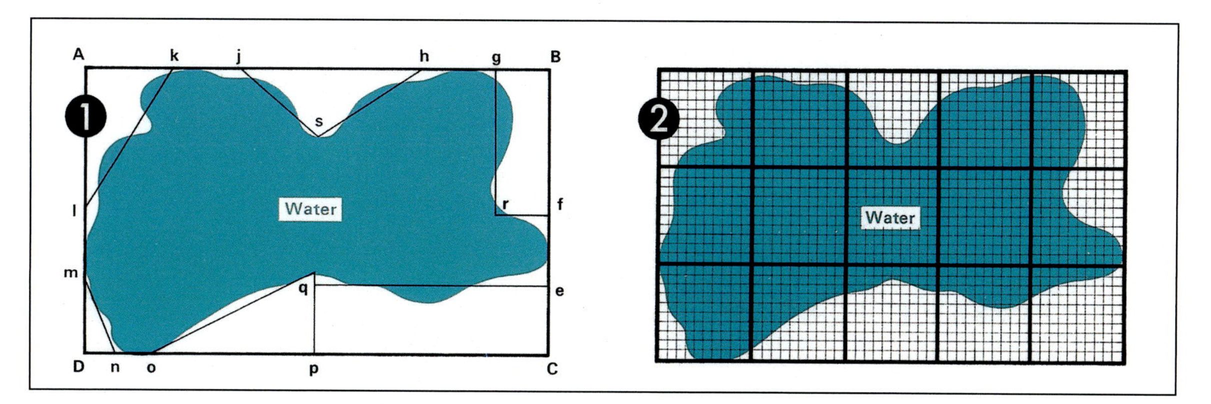

Figure 3 Method of calculating the area of an irregular shape. (1) by calculation of approximate areas. (2) by the use of squared paper.

The area of the rectangle should be calculated and the sum of the areas of the external triangles and rectangles subtracted.

More accurate results can be obtained by drawing the shape of the pond or lake to an appropriate scale upon squared paper (Figure 3(2)) and then subtracting the areas of the squares left outside from the total squared area. When adding up the squares, any which are less than one half of a square should be disregarded.

A.3.3 Area of circles

Calculation of the area of a circle involves the use of the constant pi which is represented by the Greek symbol π. It is the ratio of the circumference (C) of a circle to its diameter (D). Careful measurement of both on any circular cylinder, as shown in Figure 4, will show that:

$$\pi = \frac{C}{D} = 3.1416 \text{ or approximately } 3^1/_7$$

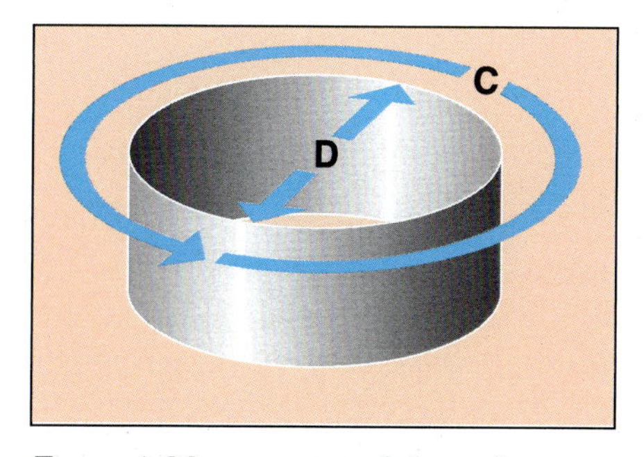

Figure 4 Measurement of circumference and diameter of a cylinder.

It can be shown that the area (A) of a circle is found by squaring the radius (r) and multiplying by π,

$$\text{i.e.} \quad A = \pi r^2$$

In hydraulic calculations, it is often more convenient to use the diameter of a circle rather than its radius so that substituting D/2 for r we have:

$$A = \pi \times \frac{D}{2} \times \frac{D}{2} = \frac{\pi D^2}{4}$$

dividing 3.1416 by 4 $= 0.7854\ D^2$

Thus, we may choose whichever is the most convenient formula for the area of a circle from:

$$A = \pi r^2 \text{ or } \frac{\pi D^2}{4} \text{ or } 0.7854\ D^2$$

The result of the calculation will be in square metres or square millimetres depending on whether the radius and diameter are expressed in metres or millimetres.

A.3.4 Volumes

When working out the area of a figure, two lengths are multiplied. In calculating volumes, three lengths must be multiplied. For containers of constant cross section this means multiplying the surface area by the depth. For large containers the three lengths are most conveniently measured in metres and the volume calculated will then be in cubic metres. Since there are 1000 litres in a cubic metre, the result of the calculation can be expressed in litres simply by multiplying by 1000.

(a) Rectangular tanks

The volume of a rectangular tank (Figure 5(1)) is calculated by multiplying the length (l) by the breadth (b) by the depth (h).

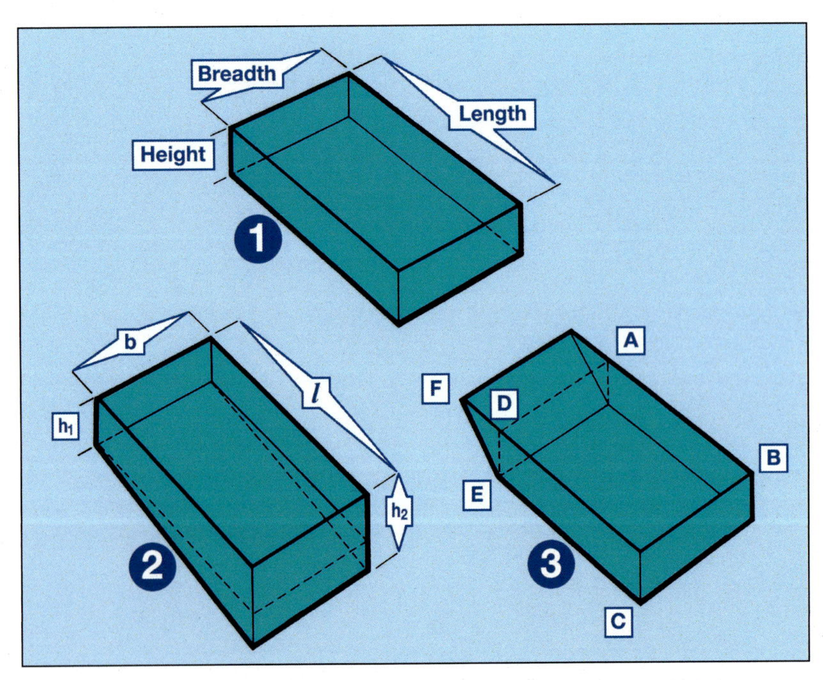

Figure 5 Volumes of (1) a rectangular tank; (2) a rectangular tank with a uniformly sloping base and (3) a rectangular tank with a sloping end.

If the dimensions of the tank in Figure 5(1) are length 7.5 metres, breadth 2 metres, depth 1 metre, the volume (in cubic metres) will be:

$$7.5 \times 2 \times 1 = 15 \text{ cubic metres } (m^3) \text{ or } 15 \times 1000 = 15\,000 \text{ litres}$$

Capacity of a rectangular tank in litres = $l \times b \times h \times 1000$

(b) Rectangular tanks with a sloping base

The capacity of rectangular tanks with uniformly sloping bases such as swimming pools, can be obtained by proceeding as in (1) above but multiplying by the average depth, which is ascertained by adding together the values for the deep and shallow ends and dividing by 2. In Figure 5(2) the capacity (all dimensions being in metres) would be:

$$\text{capacity} = l \times b \times \frac{h_1 + h_2}{2} \text{ (cubic metres)}$$

Where the container is more complex in shape it is usually possible to estimate its total capacity with a reasonable degree of accuracy by dividing it up into a number of simpler shapes for each of which it is easy to calculate the capacity. For example Figure 5(3) shows a tank with a sloping end. The capacity of such a tank is easily calculated by dividing it into a rectangular part ABCD with uniform depth DE, and a triangular part DEF.

The volume of a tank of triangular section is obtained by multiplying the area of the triangle DEF by the length (b).

Therefore, the capacity of the tank shown in Figure 5(3) is found by adding together the capacities of the rectangular and triangular portions.

(c) Circular tanks

The volume of a circular tank (Figure 6) is determined by calculating the surface area and multiplying by the depth (h), all measurements being in the same units. This can be expressed as

$$\pi r^2 h \quad \text{or} \quad \frac{\pi D^2 h}{4} \quad \text{or} \quad 0.7854\, D^2 h$$

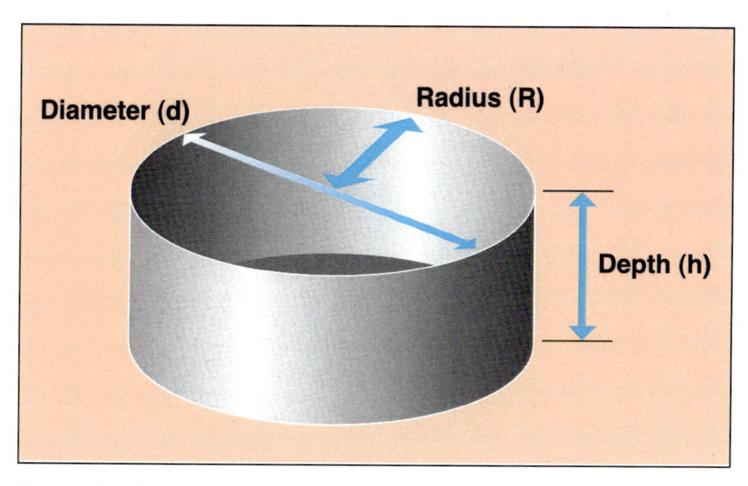

Figure 6 Volume of a circular tank.

The capacity of a circular tank 10 metres in diameter and 4 metres deep would be:

$$0.7854 \times 10 \times 10 \times 4 = 314.16 \text{ cubic metres}$$

Multiplying by 1000 gives the capacity in litres = 314 160 litres.

(1) Quick method 1

The capacity of a circular tank has been shown to be $0.7854 \times D^2h$ cubic metres, but

$$0.7854 \simeq 0.8$$

therefore a good approximation is:

Quick formula 1: Capacity of circular tank = $0.8\ D^2h$ cubic metres (m^3)

where D = diameter and h = depth, both in metres.
In the above example, this would give

$$0.8 \times 10 \times 10 \times 4 = 320 \text{ cubic metres}$$
(which is about 2 per cent too high)

Alternatively, the capacity in litres can be obtained by multiplying by 1000

i.e. **Capacity of circular tank = $800\ D^2h$ litres**

(2) Quick method 2

The capacity of a circular tank when only the circumference, (C), and depth, (d), are known is found by making the substitution:

$$D = \frac{C}{\pi}$$

in the formula:

$$\text{capacity} = \frac{\pi D^2h}{4}$$

giving:

$$\text{capacity} = \frac{\pi}{4} \times \frac{C}{\pi} \times \frac{C}{\pi} \times h$$

$$= \frac{C^2h}{4\pi}$$

$$= \frac{C^2h}{12.5664}$$

As 12.5664 is approximately $= \frac{100}{8}$

another quick formula for calculating the capacity of a circular tank in cubic metres is:

Quick formula 2: capacity of circular tank

$$= \frac{8C^2h}{100} \quad \textbf{cubic metres}$$

or, multiplying by 1000 to convert cubic metres to litres:

$$= 80\ C^2h \quad \textbf{litres}$$

The error introduced when either of these two formulae is used is only about 0.5% and is therefore likely to be within the limits of accuracy of the measurements themselves when these have been obtained in a practical situation.

(d) Capacity of hose or pipeline

Once a length of hose is filled with water under pressure it becomes a horizontal circular cylinder so one of the approximate formulae derived above, such as:

$$\text{capacity} = 800\ D^2h \quad \textbf{litres}$$

may be applied. In this formula D is the diameter in metres and h is now the hose *length.* However, hose diameter is usually measured in millimetres so we must make the substitution:

$$D = \frac{d}{1000}$$

giving:

$$\text{capacity} = 800 \times \frac{d}{1000} \times \frac{d}{1000} \times h \text{ litres}$$

$$= \frac{8\ d^2\ h}{10\ 000} \quad \text{litres}$$

Taking just 1 metre of hose (i.e. h = 1) gives:

$$\textbf{capacity} = \frac{8\ d^2}{10\ 000} \quad \textbf{litres per metre}$$

(e) Cones, pyramids and spheres

Firefighters are not often required to calculate the volume of cones, pyramids or spheres when dealing with hydraulic problems, yet it is advantageous to know the formulae for such calculations.

(1) Cones and pyramids

The volume of a cone or pyramid (Figure 7) is obtained by multiplying the area of the base by one-third of the vertical height.

$$\text{Volume of a cone or pyramid} = \frac{\text{area of base} \times h}{3}$$

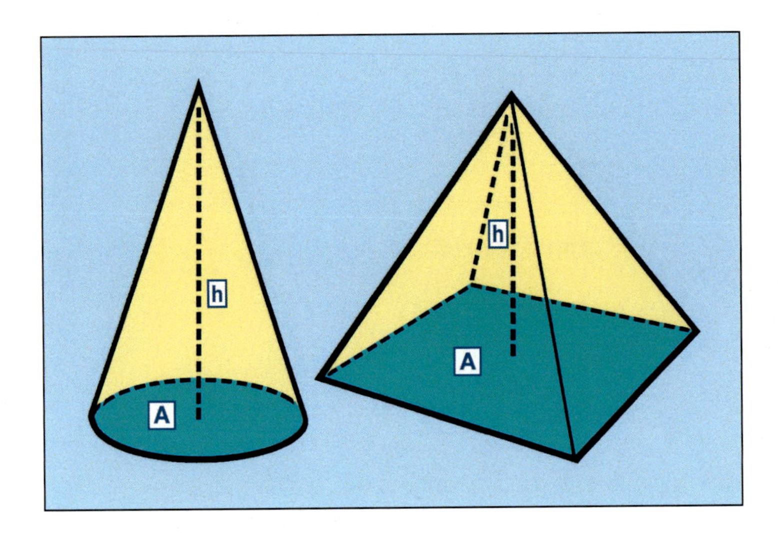

Figure 7 Volume of a cone or pyramid.

(2) *Spheres*

To calculate the volume of a sphere, the cube of the diameter is multiplied by $\pi/6$, or the cube of the radius is multiplied by $4\pi/3$

$$\textbf{Volume of a sphere} = \frac{\pi d^3}{6} \text{ or } \frac{4 \pi r^3}{3}$$

(f) Irregular-shaped tanks

The volume or capacity of irregular-shaped tanks or ponds is calculated by finding the surface area by one of the methods already explained and multiplying by the depth. In the majority of cases the depth of ponds and lakes is not uniform, and the capacity is conservatively calculated as being two-thirds of the value obtained by multiplying the surface area by the average depth.

$$\textbf{Capacity of a pond or lake (cubic metre)} = \frac{2}{3}\,\textbf{(surface area} \times \textbf{average depth)}$$

A very approximate estimate of the capacity of such a pond or lake in litres would be 700 × surface area in square metres × average depth in metres.

Derivation of Hydraulics Formulae

The derivations which follow assume an understanding of the concepts of **force, work, energy, power** etc. many of which are discussed in the Manual of Firemanship Volume 1 'Physics and Chemistry for Firefighters'.

A.4.1 Relationship between Pressure and Head

Figure 1 shows a vessel having an area of cross section 1 square metre and height H metres filled with a liquid of density ρ kilograms per cubic metre.

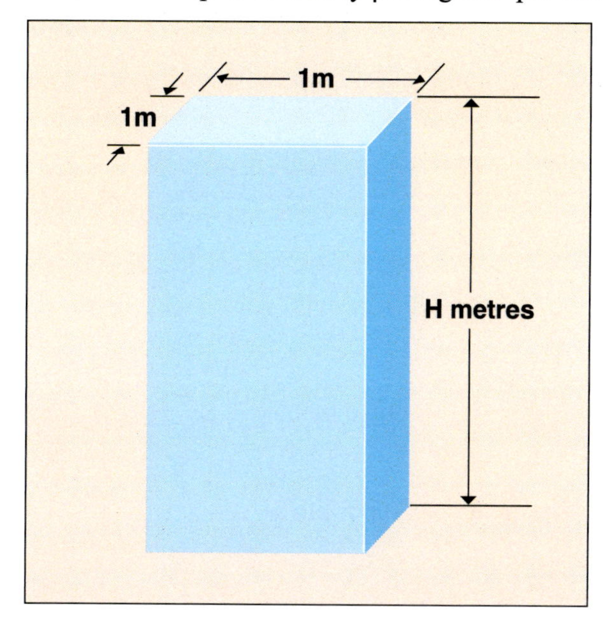

Figure 1

The volume of liquid in the container is:

$1 \times H$ cubic metres

and its mass (volume x density) is:

$H\rho$ kilograms

Remembering that each kilogram of mass experiences a force due to gravity of g newtons, the weight of this column of liquid (and hence the force exerted on the base of the container) is:

$H\rho g$ newtons

Because pressure, p, is defined as the force acting on unit area then

$\mathbf{p = H\rho g}$ newtons per square metre

A.4.2 Velocity and Flowrate in Hose and Pipes

Figure 2 shows a liquid moving with a velocity of v metres per second in a pipe of diameter d millimetres. In 1 second the volume of liquid discharged will be that of a circular cylinder v metres long and d millimetres in diameter.

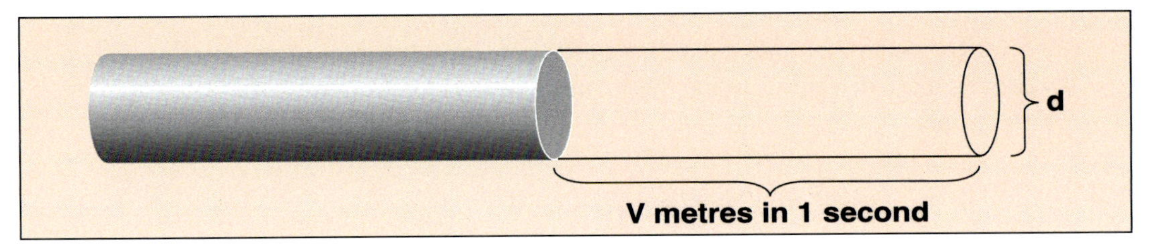

Figure 2

The volume of a circular cylinder is given by the formula:

$$0.7854\,D^2h \qquad \text{cubic metres}$$

where D is the diameter in metres and h is the height or length in metres.

Making the substitutions:

$$D = \frac{d}{1000} \text{ and } h = v$$

$$\text{Volume discharged in 1 second} = 0.7854 \times \frac{d}{1000} \times \frac{d}{1000} \times v \qquad \text{cubic metres}$$

$$= 0.7854 \times \frac{d}{1000} \times \frac{d}{1000} \times v \times 1000 \text{ litres}$$

$$= \frac{0.7854vd^2}{1000} \qquad \text{litres}$$

The number of litres discharged in 1 minute is therefore:

$$L = \frac{60 \times 0.7854vd^2}{1000} \qquad \text{litres}$$

i.e.

$$\mathbf{L = \frac{vd^2}{21.2}} \qquad \textbf{litres per minute}$$

The transposed version of this formula:

$$\mathbf{v = \frac{21.2L}{d^2}} \qquad \textbf{metres per second}$$

is useful for finding the velocity of flow in a hose or pipe when the flowrate is known.

A.4.3 Water Power (WP)

Figure 3 shows a pump which is discharging L litres of water per minute at a pressure of P bars along a pipe with an area of cross section of A square metres. Suppose the distance moved by the column of water in 1 minute is *l* metres.

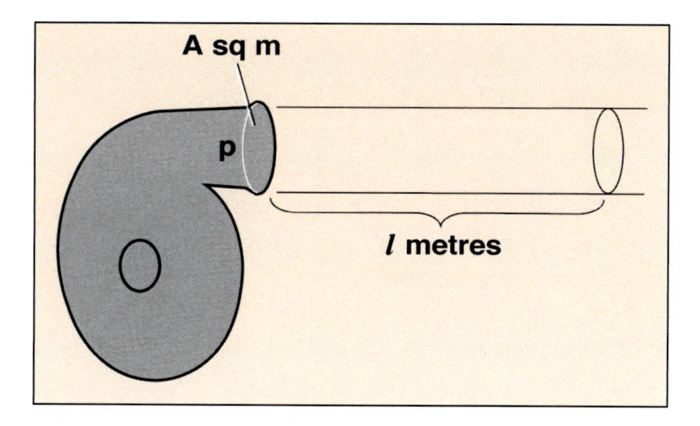

Figure 3

The **force** (pressure × area) exerted on the column of water is:

$$100\,000 \times P \times A \text{ newtons (N)}$$

(the factor of 100 000 appears because 1 bar = 100 000 N/m^2)

The **work** (force × distance) done by the pump in 1 minute is:

$$100\,000 \times P \times A \times l \text{ joules (J)}$$

But A × *l* is the **volume** of water discharged in 1 minute measured in cubic metres and this, in turn, is equal to the number of litres, L, discharged in 1 minute divided by 1000.

So, the **work** done by the pump in 1 minute is:

$$100\,000 \times P \times \frac{L}{1000} \text{ joules}$$

$$= 100LP \text{ joules}$$

The **water power** (power is defined as energy expended or work done per *second*) delivered by the pump is therefore given by the formula:

$$\mathbf{WP = \frac{100LP}{60} \text{ watts (W)}}$$

Strictly speaking P is the **increase** in pressure created by the pump so, if there is a positive pressure at the inlet, this value should be **subtracted** from the outlet pressure. Likewise, if the inlet pressure is negative because the pump is lifting from an open source, the appropriate value should be **added** to the outlet pressure to take into account the work the pump is doing in lifting water.

A.4.4 Bernoulli's formula

This is a relationship involving the pressure, velocity and height of a liquid flowing along a pipe of variable diameter at two different points in the pipe. Its derivation makes use of one of the fundamental principles of science; the Principle of the Conservation of Energy. This states that energy may neither be created nor destroyed but only transformed from one form into another.

The formula is included here because many important concepts of hydraulics follow directly from it.

Figure 4 shows a liquid flowing along a pipe between two points indicated in the figure as position 1 and position 2. At position 1 the area of cross section is A_1 square metres, the pressure p_1 newtons per square metre, the velocity v_1 metres per second and the height above some arbitrary horizontal reference line (e.g. the ground) h_1 metres. At position 2 these values change to, respectively, A_2, p_2, v_2, and h_2.

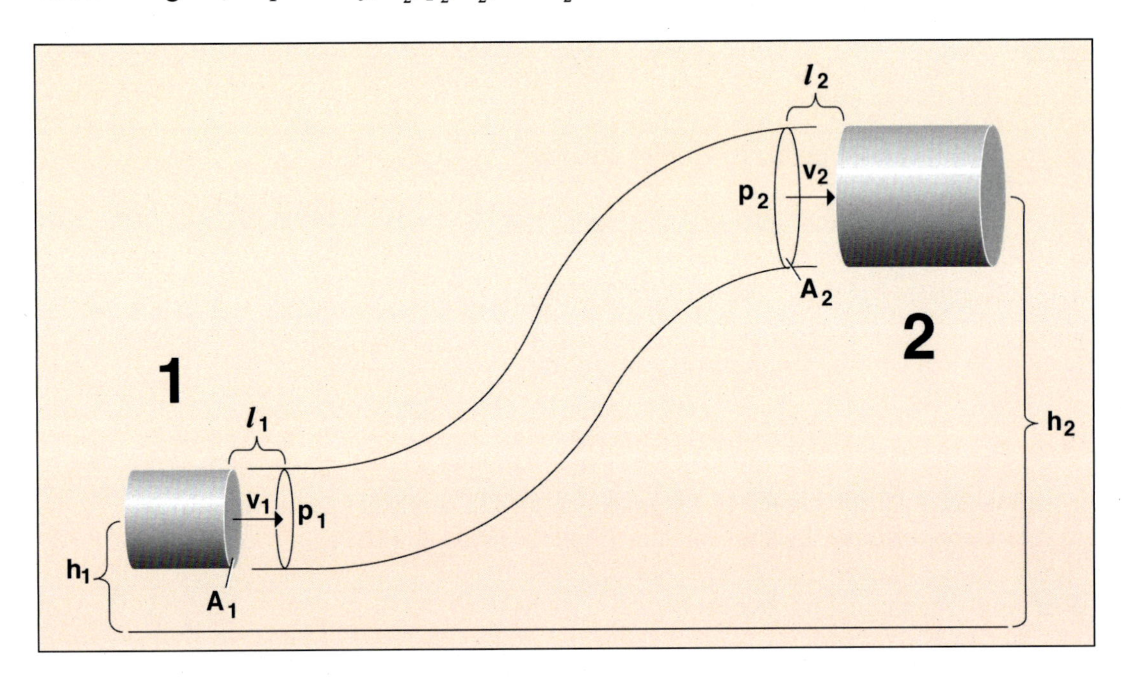

Figure 4

Suppose the liquid at position 1 travels a short distance l_1 along the pipe and that, in the same period of time, liquid at position 2 travels a distance l_2. If we assume that the liquid is incompressible then the two volumes displaced, and consequently the two masses, will be the same.

The two volumes displaced are:

$$A_1 l_1 \quad \text{and} \quad A_2 l_2$$

and the two masses (volume × density):

$$A_1 l_1 \rho \quad \text{and} \quad A_2 l_2 \rho$$

Applying the principle of the conservation of energy:

Total energy of the liquid at position 1 + work done by the pressure p_1 =
Total energy of liquid at position 2 + work done by the pressure p_2

The energy at both positions is made up of potential energy (mgh) and kinetic energy $\left(\frac{mv^2}{2}\right)$

The work done by the pressure at position 1 is:

force × distance
= pressure × area × distance
$= p_1A_1l_1$

and likewise at position 2 $= p_2A_2l_2$

Thus

$$\frac{mv_1^2}{2} + mgh_1 + p_1A_1l_1 = mv_2^2 + \frac{mgh_2}{2} + p_2A_2l_2$$

Dividing throughout by m gives:

$$\frac{v_1^2}{2} + gh_1 + \frac{p_1A_1l_1}{m} = \frac{v_2^2}{2} + gh_2 + \frac{p_2A_2l_2}{m}$$

Substituting $A_1l_1\rho$ and $A_2l_2\rho$ for m on the left- and right-hand side respectively:

$$\frac{v_1^2}{2} + gh_1 + \frac{p_1A_1l_1}{A_1l_1\rho} = \frac{v_2^2}{2} + gh_2 + \frac{p_2A_2l_2}{A_2l_2\rho}$$

giving:

$$\mathbf{\frac{v_1^2}{2} + gh_1 + \frac{p_1}{\rho} = \frac{v_2^2}{2} + gh_2 + \frac{p_2}{\rho}}$$

In this form of Bernoulli's equation, the three terms on each side of the equals sign represent, respectively, the kinetic energy, potential energy and pressure energy per kilogram of liquid. The large number of variables in the formula make it appear complex but, in many applications, it will simplify because certain factors take the same value on both the left- and right-hand sides of the equals sign. For example, if we apply the formula to a nozzle, both the inlet and the outlet are, for all practical purposes, at the same horizontal level so that the two gh terms are the same and the formula reduces to:

$$\frac{v_1^2}{2} + \frac{p_1}{\rho} = \frac{v_2^2}{2} + \frac{p_2}{\rho}$$

Before attempting to use the Bernoulli formula it should be noted that pressure must be measured in newtons per square metre (pascals) and that, for water, the density, ρ, may be taken as 1000kg per cubic metre.

An alternative version of the formula expresses the pressures at the two points in the pipe in terms of the head in metres at those points. It is obtained by making the substitutions:

$$p_1 = H_1\rho g \quad \text{and} \quad p_2 = H_2\rho g$$

Thus:

$$\frac{v_1^2}{2} + gh_1 + \frac{H_1\rho g}{\rho} = \frac{v_2^2}{2} + gh_2 + \frac{H_2\rho g}{\rho}$$

i.e.

$$\frac{v_1^2}{2} + gh_1 + H_1g = \frac{v_2^2}{2} + gh_2 + H_2g$$

Dividing throughout by g gives:

$$\mathbf{\frac{v_1^2}{2g} + h_1 + H_1 = \frac{v_2^2}{2g} + h_2 + H_2}$$

Again, if the pipe is horizontal so that h_1 and h_2 are the same, this simplifies to give:

$$\frac{v_1^2}{2g} + H_1 = \frac{v_2^2}{2g} + H_2$$

A.4.5 Velocity of Jets

Figure 5 shows a conventional straight stream nozzle operating at a pressure of P bars and discharging water at a velocity of v metres per second.

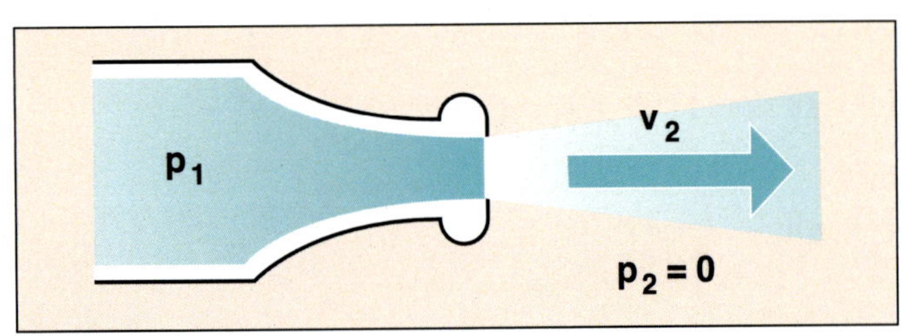

Figure 5

Since water enters and leaves the nozzle at the same horizontal level, then we may apply the simplified version of Bernoulli's formula:

$$\frac{v_1^2}{2} + \frac{p_1}{\rho} = \frac{v_2^2}{2} + \frac{p_2}{\rho}$$

If the hose diameter is large compared with the nozzle diameter, then the velocity, v_1, of the water when it is in the hose may, without undue error, be taken as zero. Also, the (gauge) pressure, p_2, of the water as it leaves the nozzle is zero. Making these two substitutions gives:

$$\frac{p_1}{\rho} = \frac{v_2^2}{2}$$

Because 1 bar = 100 000 newtons per square metre (i.e. p_1= 100 000P) and ρ, the density of water, is 1000kg per cubic metre, then:

$$\frac{100\,000P}{1000} = \frac{v_2^2}{2}$$

i.e. $$v^2 = 200P$$

or $$\mathbf{v = 14.14\sqrt{P}}$$

A.4.6 The Nozzle Discharge Formula

This formula is obtained by substituting the expression for v given by the velocity of jets formula into the relationship for flowrate (L), in terms of pipe diameter (d) and velocity of flow (v):

$$L = \frac{vd^2}{21.2}$$

i.e. $$L = \frac{14.14\sqrt{P}\,d^2}{21.2}$$

i.e. $$\mathbf{L = \frac{2d^2\sqrt{P}}{3}}$$

A.4.7 The Jet Reaction Formula

Figure 6 shows water discharging with a velocity v metres per second from a conventional straight stream nozzle of diameter d millimetres and operating at a pressure of P bars.

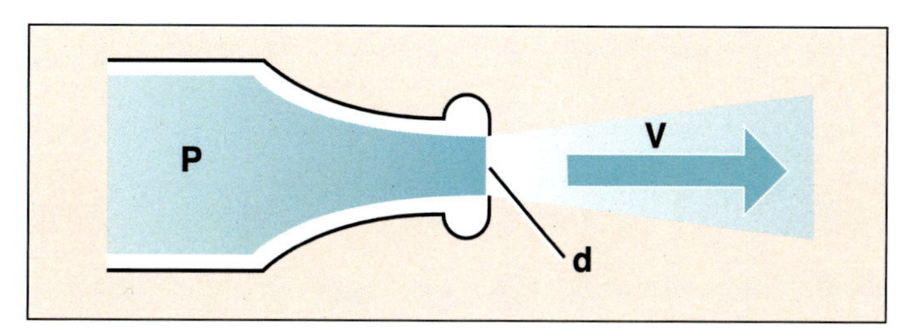

Figure 6

To obtain the formula for jet reaction we make use of the relationship, discussed in the Manual of Firemanship Volume 1 'Physics and Chemistry for Firefighters', between **force, mass** and **acceleration.**

i.e. **force = mass × acceleration**

Where the force, F, is measured in newtons, the mass, m, being accelerated is measured in kilograms and the resulting acceleration, a, is measured in metres per second per second.

Acceleration is defined as change in velocity in 1 second. For a body starting from rest it becomes acquired velocity, v, divided by time taken, t.

So $$F = m\frac{v}{t}$$

We can apply the above formula to the water discharged from a straight stream nozzle in 1 minute. In this case m is the mass in kilograms discharged in 1 minute which is numerically equal to the number of litres, L, so:

APPENDIX 4 continued

$$F = \frac{Lv}{t}$$

But: $L = \frac{2d^2\sqrt{P}}{3}$, $v = 14.14\sqrt{P}$ and $t = 60$

so $F = \frac{2d^2\sqrt{P} \times 14.14\sqrt{P}}{3 \times 60}$

$= 0.157\ P\ d^2$

This formula gives the force which must be applied to the water in the hose in order to create the jet but, according to Newton's third law of motion (action and reaction are equal and opposite), it is the same as the reaction experienced by whoever or whatever is supporting the branch. Thus:

$$\mathbf{R = 0.157\ P\ d^2}$$

The reaction, **R**, is measured in newtons, the nozzle pressure, **P**, in bars and the nozzle diameter, **d**, in millimetres.

The reaction may also be expressed in terms of the area of cross section, **a**, of a nozzle:

Since $a = \frac{3.14\ d^2}{4}$ sq. mm

then $d^2 = \frac{4a}{3.14}$

so $R = \frac{0.157P \times 4a}{3.14}$

i.e. $\mathbf{R = 0.2Pa}$

where **a** is the area of cross section of the nozzle in square millimetres.

Summary of Formulae and Other Data

A.5.1 Approximate fireground calculation

Loss of pressure due to height = 0.1 bar for each metre rise

Capacity of pond or lake (m^3) $= \frac{2}{3}$ (surface area (m^2) × average depth (m))

Capacity of circular tank (m^3) $= 0.8D^2h$

$= \frac{8C^2h}{100}$

In litres: $= 800D^2h$

$= 80\ C^2h$

A.5.2 Hydraulic formulae

Capacity of hose (litres/m) $= \frac{8d^2}{10\,000}$ (approx)

Pressure and head

$P = H \times 0.0981$

$H = P \times 10.19$

$P = \frac{H}{10}$ (approx)

$H = P \times 10$ (approx)

Water power (watts) $WP = \frac{100 \times L \times P}{60}$

Percentage efficiency $E = \frac{WP}{BP} \times 100$

Brake power (watts) $BP = \frac{WP \times 100}{E}$

Velocity and discharge $v = \frac{21.2L}{d^2}$

$L = \frac{vd^2}{21.2}$

Friction loss $P_f = \frac{9000flL^2}{d^5}$

Nozzle discharge $L = \frac{2\, d^2\sqrt{P}}{3}$

Jet reaction $R = 0.2Pa$

$= 0.157Pd^2$

A.5.3 Hydraulic data

1 litre of water has a mass of	1 kilogram
1 litre of water exerts a downward force of	approx 10 newtons (N)
1 cubic metre of water exerts a downward force of	approx 10 kN
1 metre head of water equals	approx 0.1 bar
1 bar pressure of water equals	approx 10 metres head

A.5.4 Constants

g (acceleration due to gravity)	$= 9.81\ m/s^2$
Normal atmospheric pressure at 20°C	= 1.013 bar
Normal atmospheric pressure at 20°C	= 10.3 metre head of water
π (pi)	$= 3.1416\ (3^1/_7)$
$\frac{\pi}{4}$	= 0.7854
Circumference of a circle	$= \pi d$ (or) $2\pi r$

A.5.5 Areas

Circle	$= \pi r^2$ (or) $\frac{\pi d^2}{4}$
Triangle (s = ½ sum of sides)	$= \sqrt{s \times (s-a)(s-b)(s-c)}$
	$= \frac{\text{base} \times \text{perp height}}{2}$

A.5.6 Volumes

Sloping tank	= length × breadth × average depth
Circular tank (cylinder)	$= \pi r^2 \times \text{depth or } \frac{\pi d^2}{4} \times \text{depth}$
Cone or pyramid	$= \frac{\text{Area of base} \times \text{vert. height}}{3}$
Sphere	$= \frac{\pi d^3}{6} \text{ or } \frac{4\pi r^3}{3}$

A.5.7 Capacity measured in litres

Capacity of a container in litres	= Volume (cu. metres) × 1000

Sections 57 and 58 of the Water Industry Act 1991

"57.1 It shall be the duty of a water undertaker to allow any person to take water for extinguishing fires from any of its water mains or other pipes on which a fire hydrant is fixed.

*57.2 Every water undertaker shall, at the request of the fire authority concerned, fix fire hydrants on its water mains (other than its trunk mains) at such places as may be convenient for affording a supply of water for extinguishing any fire which may break out within the area of the undertaker.

57.3 It shall be the duty of every water undertaker to keep every fire hydrant fixed on any of its water mains or other pipes in good working order and, for that purpose, to replace any such hydrant when necessary.

57.4 It shall be the duty of a water undertaker to ensure that a fire authority has been supplied by the undertaker with all such keys as the authority may require for the fire hydrants fixed on the water mains or other pipes of the undertaker.

57.5 Subject to section 58.3 below, the expenses incurred by a water undertaker in complying with its obligations under subsections 2 to 4 above shall be borne by the fire authority concerned.

57.6 Nothing in this section shall require a water undertaker to do anything which it is unable to do by reason of the carrying out of any necessary works.

57.7 The responsibilities of a water undertaker under this section shall be enforceable under section 18 above by the Secretary of State.

57.8 In addition, where a water undertaker is in breach of its obligations under this section, the undertaker shall be guilty of an offence . . .

* *In spite of this clause, some water authorities are prepared to allow hydrants to be fixed on trunk mains.*

57.9 In any proceedings against any water undertaker for an offence under subsection 8 above it shall be a defence for that undertaker to show that it took all reasonable steps and exercised all due diligence to avoid the commission of the offence."

Section 58 of the Act relates to specially requested fire hydrants at factories or places of business and reads as follows:

"58.1 A water undertaker shall, at the request of the owner or occupier of any factory or place of business, fix a fire hydrant, to be used for extinguishing fires and not other purposes, at such place on any suitable water main or other pipe of the undertaker as near as conveniently possible to that factory or place of business.

58.2 For the purpose of subsection 1 above a water main or other pipe is suitable, in relation to a factory or place of business, if –

(a) it is situated in a street which is in or near to that factory or place of business; and

(b) it is of sufficient dimensions to carry a hydrant and is not a trunk main.

58.3 Subsection 5 of section 57 above should not apply in relation to expenses incurred in compliance, in relation to a specially requested fire hydrant, with its obligations under subsections 3 and 4 of that section.

58.4 Any expenses incurred by a water undertaker –

(a) in complying with its obligations under subsection 1 above; or

(b) in complying, in relation to a specially requested fire hydrant, with its obligations under section 57(3) and 57(4) above,

shall be borne by the owner or occupier of the factory or place of business in question, according to whether the person who made the original request for the hydrant did so in their capacity as owner or occupier."

APPENDIX 7

Metrication

List of SI units for use in the fire service.

Quantity and basic or derived SI unit and symbol	Approved unit of measurement	Conversion factor
Length metre (m)	kilometre (km) metre (m) millimetre (mm)	1 mile = 1.609 km 1 yard = 0.914m 1 foot = 0.305m 1 inch = 25.4 mm
Area square metre (m^2)	square kilometre (km^2) square metre (m^2) square millimetre (mm^2)	1 $mile^2$ = 2.590 km^2 1 $yard^2$ = 0.836 m^2 1 $foot^2$ = 0.093m^2 1 $inch^2$ = 645.2 mm^2
Volume cubic metre (m^3)	cubic metre (m^3) litre (l) (= $10^{-3}m^3$)	1 cubic foot = 0.028 m^3 1 gallon = 4.546 litres
Volume, flow cubic metre per second (m^3/s)	cubic metre per second (m^3/s) litres per minute (l/min = $10^{-3}m^3$/min)	1 $foot^3$/s = 0.028 m^3/s 1 gall/min = 4.546 l/min
Mass kilogram (kg)	kilogram (kg) tonne (t) (1 tonne = 10^3kg)	1 lb = 454 kg 1 ton = 1.016 t
Velocity metre per second (m/s)	metre/second (m/s) International knot (kn) kilometre/hour (km/h)	1 foot/second = 0.305 m/s 1 Int. knot = 1.852 km/h 1 UK knot = 1.853 km/h 1 mile/hour = 1.61 km/h
Acceleration metre per second² (m/s^2)	metre/second²	1 foot/second² = 0.305 m/s^2 'g' = 9.81 m/s^2
Force Newton (N)	kiloNewton (kN) Newton (N)	1 ton force = 9.964 kN 1 lb force = 4.448 N

APPENDIX 7

Quantity and basic or derived SI unit and symbol	Approved unit of measurement	Conversion factor
Energy, work Joule (J) (= 1 Nm)	joule (J) kilojoule (kJ) kilowatt-hour (kWh)	1 British thermal unit = 1.055 kJ 1 foot lb force = 1.356 J
Power watt (W) (= 1 J/s = 1 Nm/s)	kilowatt (kW) watt (W)	1 horsepower = 0.746 kW 1 foot lb force/second = 1.356W
Pressure newton/metre^2 (N/m^2)	bar = 10^5 N/m^2 millibar (m bar) (= 10^2 N/m^2) metrehead	1 atmosphere = 101.325 kN/m^2 = 1.013 bar 1 lb force/in^2 = 6894 76 N/m^2 = 0.069 bar 1 inch Hg = 33.86 m bar 1 metrehead = 0.0981 bar 1 foothead = 0.305 metrehead
Heat, quantity of heat Joule (J)	joule (J) kilojoule (kJ)	1 British thermal unit = 1.055 kJ
Heat flow rate watt (W)	watt (W) kilowatt (kW)	1 British thermal unit/ hour = 0.293 W 1 British thermal unit/ second = 1.055 kW
Specific energy, calorific value, specific latent heat joule/kilogram (J/kg)	kilojoule/kilogram (kJ/kg) kilojoule/m^3 (kJ/m^3) joule/m^3 (J/m^3) megajoule/m^3 (MJ/m^3)	1 British thermal unit/ lb = 2.326 kJ/kg 1 British thermal unit/ft^3 = 37.26 kJ/m^3
Temperature degree Celsius (°C)	degree Celsius (°C)	1 degree centigrade = 1 degree Celsius

Hydraulics, Pumps and Water Supplies

Acknowledgements

HM Fire Service Inspectorate is indebted to all who helped with the provision of information, expertise and validation to assist production of this manual, in particular:

Keith Davies
The Fire Service College
Martin Fraser
Tony Barnes

Fire Experimental Unit:	Dr Martin Thomas and staff
CACFOA :	CFO Martin Chapman
	CFO Peter Coombs

Graham Dash
Angus Fire
Hale Products (Godiva)
Wessex Water
Water UK

Fire Brigades:

Cheshire
Dorset
Durham and Darlington
Fife
Greater Manchester
Hampshire
Hertfordshire
London
Lothian and Borders
Merseyside
Norfolk
Shropshire
West Midlands